LOVING INSECTS

Loving INSECTS

John Abbot's Drawings & Natural History Collecting *in the* Atlantic World, 1760s–1840s

BETH FOWKES TOBIN

The University of Georgia Press

ATHENS

A Wormsloe
FOUNDATION
nature book

Athens, Georgia 30602
www.ugapress.org

Designed by Isaac Tobin
Set in Adobe Caslon by Isaac Tobin
Printed and bound by Friesens

The paper in this book meets the guidelines for permanence and durability of the Committee on Production Guidelines for Book Longevity of the Council on Library Resources.

Printed in Canada

26 27 28 29 30 C 5 4 3 2 1

EU Authorized Representative
Easy Access System Europe—Mustamäe tee 50, 10621 Tallinn, Estonia, gpsr.requests@easproject.com

Library of Congress Cataloging-in-Publication Data
Names: Tobin, Beth Fowkes author
Title: Loving insects : John Abbot's drawings and natural history collecting in the Atlantic world, 1760s–1840s / Beth Fowkes Tobin.
Description: Athens : The University of Georgia Press, [2026] | Series: Wormsloe Foundation nature books | Includes bibliographical references and index.
Identifiers: LCCN 2025045126 | ISBN 9780820373577 hardback
Subjects: LCSH: Abbot, John, 1751–1840 | Entomologists—United States—Biography | Natural history illustrators—United States—Biography | Insects—United States—History—19th century | Insects in art | Natural history illustration—History—19th century | Art and science—History—19th century | LCGFT: Biographies
Classification: LCC QL31.A25 T63 2026
LC record available at https://lccn.loc.gov/2025045126

For my family

Contents

Preface & Acknowledgments

FOR THE PAST TWO DECADES, I have studied eighteenth- and nineteenth-century natural history, including its arts and cultures of collecting. My long-standing fascination with the history of natural history began with an interest in how the natural world was represented in eighteenth-century English nature poetry. While teaching James Thomson's *The Seasons*, I became intrigued by the poem's portrayal of tropical nature, which led me to study how the tropics were depicted in eighteenth-century British travel narratives and natural histories. My book *Colonizing Nature: The Tropics in British Arts and Letters, 1760–1830* was the result of these efforts. Working on the last chapter of that book, which was on Sir James Edward Smith's *Exotic Botany*, took me to the Linnean Society in London and its library, where I read in Smith's letters about his plan to tap various botanical collectors for samples from their gardens and the frustrations he experienced as the costs of publication grew, forcing him to stop publishing this labor of love. Reading these letters at the Linnean Society, I became fascinated with naturalists' correspondence, in particular the way in which they formed networks that spanned the globe. They used these networks to exchange information, specimens, drawings, and books and created communities of like-minded people who helped each other (usually) to achieve their goals. Since that moment in the Linnean Society library, I have become obsessed with eighteenth-century natural history collectors.

Having written T*he Duchess's Shells*, a book on eighteenth-century shell collecting, I somehow thought I might be in a position to explore insect collectors, since this duchess also collected insects. But I quickly learned that insects are a whole lot more complicated than mollusks, and I needed to learn about insects before I could study those who collected insects and made drawings of them. As a result, it has taken more than ten years to research and write this book, all of which would not have been possible without the intelligence, generosity, and guidance of too many people to thank properly.

The research for this book would not have been possible without the generous support of the University of Georgia. With the support of Professor Juanita

Johnson-Bailey, former head of the Institute of Women's and Gender Studies, and Professor Michael Moran, former head of the Department of English, I applied for and was awarded a fellowship from the Office of the Senior Vice President for Academic Affairs and Provost's Study in a Second Discipline Fellows Program, which allowed me to study entomology for the academic year 2014–15. I am grateful to Dr. John L. Gittleman, former dean of the Odum School of Ecology, for hosting my studies at the school and to Dr. Byron Freeman, director of the Georgia Museum of Natural History, for welcoming me into the museum, where I had the great pleasure of meeting Dr. Joseph V. McHugh, curator of entomology and Herbert H. and Jean A. Ross Memorial Distinguished Professor of Insect Systematics and Evolutionary Biology in the Department of Entomology. He encouraged me to sit in on his Insect Taxonomy course, which turned out to be a life-changing experience. I had the privilege of joining his graduate students for their annual trip to Sapelo Island, where I observed how they went about collecting insects in a region not far from where Abbot did his collecting. Professor John Pickering was kind enough to chat with me about the moths in Georgia, and Professor James Porter met regularly with me, asking me to think about how Abbot's work impacted other naturalists and scientists. What I learned during this fellowship in a second discipline has been invaluable for my work on Abbot, and without it I would not have had the temerity to write this book.

I am also grateful to the Willson Center for Humanities and Arts of the University of Georgia for a travel grant to conduct research in libraries and museums in Britain and for a senior fellowship that gave me time to work on my book and to curate an exhibit on Abbot held at the Hargrett Rare Book and Manuscript Library. For this large exhibit, I brought together Abbot's drawings of insects and birds with specimens from the Natural History Museum on campus and student work in the art department's scientific illustration courses. I owe many thanks to Katherine M. Stein, director of the Hargrett Rare Book and Manuscript Library, and Anne Meyers Devine for asking me to put on this exhibit. They, along with Mazie Bowen, public service coordinator, and P. Toby Graham, university librarian and associate provost, supported every step of putting this exhibit together. Mary Linnemann, digital imaging coordinator, was fantastic at turning Abbot's fragile drawings on paper into amazing posters to decorate the walls of the exhibit. Richard Hoeboke, associate curator of entomology, contributed insect and spider specimens, while Nikole Castleberry, curator of ornithology, lent bird skins to accompany Abbot's drawings of birds. Every sixth grader in Athens's public schools visited the exhibit as part of their science curriculum, and I was delighted to see them racing around the exhibit, pointing at study skins of mergansers, Cooper's hawks, and barred owls, and squealing at the dead spiders in vials next to an enlarged drawing of a black widow spider.

Identifying the current scientific and common names of the insects and spiders in Abbot's drawings is extremely challenging and in many cases impossible. The majority of the North American insects and spiders Abbot drew were at that time unknown to European naturalists, and as a result, they did not have scientific names. In his notes Abbot supplied local common names when possible and scientific names of the genus or family of European insects that had been described and named within the Linnaean system. I have put Abbot's names for the insects and spiders he drew in quotation marks, and I have included Smith's scientific names for these insects in italics, also in quotation marks. I have tried to supply today's common and scientific names whenever possible; these are in parentheses, following the quoted material from Abbot and Smith. I have relied on the generosity and expertise of others to help with this process. When I was stumped by the current scientific names for Abbot's tiger moths, I was lucky enough to receive help from B. Christian Schmidt, research scientist at the Canadian National Collection of Insects, Arachnids, and Nematodes, who is at work on the taxonomic revision of the genera *Apantesis* and *Hemaris*. For moth IDs, I relied on such field guides as David L. Wagner's *Caterpillars of Eastern North America: A Guide to Identification and Natural History* (2005) and Seabrooke Leckie and David Beadle's *Peterson Field Guide to Moths of Southeastern North America* (2018). Invaluable were online databases such as the Global Biodiversity Information Facility (https://www.gbif.org/), BugGuide (https://bugguide.net/node/view/15740), the website "Featured Creatures" created by the Entomology and Nematology Department of the University of Florida (https://edis.ifas.ufl.edu/collections/series_featured_creatures), and the North American Moth Photographers Group at Mississippi State University (https://mothphotographersgroup.msstate.edu/). When I was feeling brave, I dipped into the intimidating publications of the Wedge Entomological Research Foundation.[1] No doubt, I have made mistakes and would be happy to hear from folks who know the proper scientific names for the insects Abbot drew. I learned to enjoy this process and will continue to puzzle over identifying the insects in Abbot's drawings long after this book makes it into print.

Toward the end of this project, I was fortunate enough to snag the attention of Jessica Wadleigh, collections assistant in the Division of Insects at the Field Museum in Chicago. She bravely stepped up to the challenge of identifying the insects and spiders that were way too obscure or complicated for me. I have noted in the captions those insects that she has identified. It has been fascinating watching her work through even the tiniest or fuzziest drawings to come up with the names of an insect's family, genus, and, surprisingly often, species. Her ability to identify insects that Abbot drew over two hundred years ago speaks well not only of her expertise but also of Abbot's skills as an astute observer and gifted artist.

To study the material and visual culture of eighteenth-century insect collecting requires access to the collections found in natural history museums and their libraries. Entomology curators graciously allowed me to examine their wonderful collections of insects and spiders. I am especially indebted to Jeanne Robinson and Geoff Hancock of the Hunterian in Glasgow for a tour of William Hunter's collection of insect specimens; Isabelle Charmantier, head of collections at the Linnean Society, for taking me down into the vault to see the lepidopteran specimens Abbot had collected; Suzanne Ryder, senior curator of Hymenoptera, who showed me Sir Joseph Banks's insect collection in London's Natural History Museum; and Alessandro Giusti, curator of Lepidoptera at the Natural History Museum, who let me look for specimens Abbot might have collected among the drawers containing hawkmoths; and a librarian whose name escapes me at the American Museum of Natural History in New York City who took me in the back to see Ronald S. Wilkinson's collection of antique breeding boxes.

Stored in the libraries of natural history museums are amazing paper objects: botanical and zoological drawings; illustrated natural history books; annotated natural history auction catalogs; and the manuscripts that natural history collectors generated, including their letters, journals, notebooks, diaries, accounts, and handwritten catalogs of private collections. (To capture the experience of reading these handwritten documents, I have resisted correcting their authors' spelling and grammatical errors.) The librarians at the Zoology Library, in particular Paul Martyn Cooper and Hellen Pethers, at London's Natural History Museum, where I spent several summers immersed in its marvelous collections of drawings and correspondence, were knowledgeable and encouraging and welcomed me into this awe-inspiring place. Equally gracious were the librarians and archivists of the other natural history museums I visited: the American Museum of Natural History in New York; the Carnegie Museum of Natural History in Pittsburgh, Pennsylvania; the Field Museum in Chicago; Harvard University's Museum of Comparative Zoology; Oxford University's Museum of Natural History; and the Smithsonian Institution in Washington, D.C. Visits to the Alexander Turnbull Library in Wellington, New Zealand; the library of the Royal Entomological Society in St. Albans, England; the Kenan Research Center at Atlanta's History Center; and the library of the American Philosophical Society in Philadelphia enriched my understanding of the context in which Abbot worked. The rare book rooms, archives, and special collections of university libraries were central in helping me come to grips with the breadth and depth of Abbot's contribution to the study of North American natural history, and I am grateful for the welcome I received at the libraries of the University of Connecticut, Cornell University, Emory University, Harvard University, Johns Hopkins University, the University of Massachusetts, the University of South Carolina, and, of course, the University of Georgia.

I am grateful to the museums and libraries that supplied the images for this book, often graciously waiving reproduction and publication fees: American Philosophical Society; Rose Manuscript, Archives, and Rare Book Library, Emory University; Hargrett Library, University of Georgia Libraries; Houghton Library, Harvard University; Sheridan Libraries, Johns Hopkins University; Muséum national d'Histoire naturelle, Paris; Natural History Museum, London; Irvin Department, University of South Carolina Libraries; Nationalbibliothek, Vienna; and Zentralbibliothek Zürich. I owe many thanks to those who were especially kind and helpful during this somewhat stressful process: Amy K. Kimball, the Sheridan Libraries, Johns Hopkins University; Mary Linnemann, Hargrett Library, University of Georgia Libraries; Eléa Nabat and Maud Haon Maatouk, Muséum national d'Histoire naturelle; Gertrud Oswald, Österreichische Nationalbibliothek; Elizabeth Sudduth, Irvin Department, University of South Carolina Libraries; and Angela Sun, Houghton Library, Harvard University.

I have benefited enormously from sharing my work on Abbot with audiences whose questions about and enthusiasm for his drawings helped guide my project. I am thankful to have been invited to speak in an array of settings, ranging from workshops and seminars to symposiums and academic conferences. Thanks go to the organizers of these events: Dale L. Couch, Green Symposium for the Decorative Arts, Athens, Georgia; Gaelen R. Burke, Faculty Seminar, Department of Entomology, University of Georgia; Nigel Rigby, Jordan Goodman, and Simon Werritt, "Joseph Banks: Science, Culture, and Exploration, 1743–1820," the Royal Society, London; Lucy Peltz, "Science, Self-Fashioning and Representation in Joseph Banks's Circles," the National Portrait Gallery, London; Courtney Wilder, Eighteenth-Century Studies Group, University of Michigan; Elizabeth Kowaleski-Wallace, Eighteenth-Century Seminar, Boston College; Sujata Iyengar, "Bookwork: Symposium in the History of the Book," University of Georgia; Christa Hayes and Scott Coleman, "Drawn from Nature," the Lodge on Little St. Simons Island, Georgia; and V. E. Mandrij and Giulia Simonini, "Insects and Colours Between Art and Natural History," a workshop at the Universität Konstanz, Germany. I am particularly grateful to Rachael Scarborough King, who gave me the opportunity to think about natural history in the context of manuscript culture when she included my chapter "Collecting John Abbot's Natural History Notes and Drawings" in her edited volume *After Print: Manuscript Studies and Eighteenth-Century Literature* (University of Virginia Press, 2020). I also greatly appreciate the comments from anonymous readers of my article "Drawing Insects: John Abbot and the Arts of Noticing," *Eighteenth Century* 62, no. 1 Copyright © 2021 University of

Pennsylvania Press, and the anonymous readers of this book's manuscript for the University of Georgia Press.

I owe a debt of gratitude to the many people who generously dispensed advice, encouragement, helpful information, and feedback on talks and drafts: Jordan Goodman, Deirdre Coleman, Dominik Hünniger, Leslie K. Overstreet, Julie Harvey, Stephanie Jovanovic-Kruspel, Harald Bruckner, Kay Etheridge, V. E. Mandrij, Giulia Simonini, John V. Calhoun, Craig Hanson, Elizabeth Kowaleski-Wallace, Chloe Wigston Smith, and Joan Piper. Special thanks go to the members of my writing group, Roxanne Eberle, Casie Legette, and Tricia Lootens, whose support has been constant and immeasurable. I am deeply grateful to Peggy Heard Galis, whose love of Georgia history is infectious. Her enthusiasm for Abbot sparked my own a decade ago when she took me to Southeast Georgia on a driving trip to find his will and his grave and to explore the various locations where he might have lived in the Savannah River region. Rumor has it that she carried in her car's trunk a pistol her father had given her long ago to protect her from the wild dogs that haunt the old cemeteries she is fond of exploring. Over the years when we crossed paths as neighbors, she would call out every so often, "How's that book on Abbot coming along?" which was just enough encouragement (and pressure) to keep me going. Thanks are also due to the University of Georgia Press, specifically Bethany Snead and Jon Davies, for guiding this book through the publication process, and to Mary M. Hill for excellent copyediting, and to Isaac Tobin for designing such a lovely book. I am grateful for the Wormsloe Foundation, Inc.'s generosity in supporting the publication of this beautifully illustrated book. I owe the largest debt of gratitude to my family members, who have been supportive throughout this long process, kindly listening to my chatter about insects and eighteenth-century Georgia, offering encouragement when I faced obstacles, and providing solace when I needed it, all of which allowed me to see this project through to completion during difficult times.

1

"A Peculiar Liking For Insects"

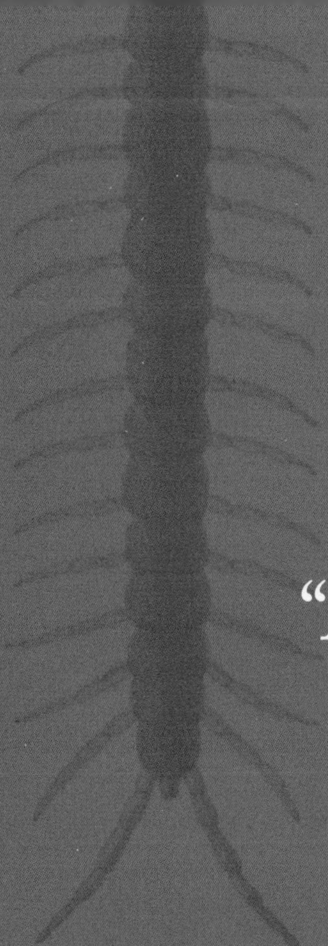

AN INTRODUCTION

FIGURE 1.1.
John Abbot, "Common American Tiger Moth (*Phalaena Phyllira*)," phyllira tiger moth (*Apantesis phyllira*), determined by J. K. Wadleigh, original drawing, TAB. LXIV, Smith and Abbot, *Insects of Georgia* (1797). Courtesy of John Work Garrett Library, Sheridan Libraries, Johns Hopkins University.

ANDREY AVINOFF (1884–1949), Russian émigré, bibliophile, and director of the Carnegie Museum of Natural History in Pittsburgh, Pennsylvania, in the 1930s and 1940s, described an album of insect drawings executed in the late 1760s and early 1770s as "an object of amazement." For Avinoff, a gifted artist and a distinguished lepidopterist, these drawings were "among the masterpieces of entomological portraiture."[1] The artist responsible for these entomological portraits was John Abbot, who was barely twenty when he drew these pictures. Born in London in 1751, he left England in 1773 to travel to the American South in search of insects to collect and draw, eventually settling in Georgia. This book is about Abbot's entomological drawings and the cultural and scientific contexts in which he produced more than six thousand drawings of North American insects and spiders.

Abbot's artwork and accomplishments as a naturalist are not widely known today. Yet Avinoff was so enamored of this young man's artwork that he chose a bound volume of Abbot's watercolor drawings to take with him when he fled his Ukrainian estate at the start of the Russian Revolution. Avinoff realized that as a member of the landed gentry, he would have to go into exile. He took care to depart in such a way as to not arouse suspicion, packing one suitcase as if he were leaving to attend a scientific conference in France. Of the thousands of items in his library, including rare natural history books such as August Johann Rösel von Rosenhof's *Der Insecten-Belustigung* (1746–61), that he could have chosen to take with him, he selected Abbot's drawings of wasps, moths, and butterflies. Avinoff wrote of this moment: "It is the *only* book I took with me in a suit case when I left Russia for good, since I considered it the most precious piece in my old library."[2]

The quality of Abbot's natural history illustrations, particularly his entomological drawings, and the knowledge he possessed as a naturalist, combined with a long and productive life, most of it lived in the American South, make Abbot a unique figure in the history of transatlantic Anglophone natural history.[3] Of the thousands of watercolor drawings Abbot drew of insects, spiders, and birds, none

of his bird drawings were engraved and printed.[4] Only 172 of his insect drawings were published, and most of these are in the magnificent two volumes of *The Natural History of the Rarer Lepidopterous Insects of Georgia*, by James Edward Smith and John Abbot, published in London in 1797. The illustrations for *Insects of Georgia* depict a butterfly or moth flitting around a cutting of its flowering host plant, images that were no doubt influenced by Maria Sibylla Merian's and Moses Harris's lepidopteran illustrations. Sir James E. Smith (1759–1828), president of London's Linnean Society, thought Abbot's drawings for this book "were the most perfect thing I ever saw."[5] Abbot was the first naturalist living in the United States to have his natural history artwork published in folio, preceding John James Audubon by two decades. Although *Insects of Georgia* was praised by one of his contemporaries as "one of the most beautiful and valuable" natural history publications of the era, Abbot's accomplishments as an artist and a naturalist are barely recognized today beyond a handful of ornithologists and lepidopterists, natural history museum curators and librarians, and collectors of natural history prints and drawings of regional significance.[6]

Abbot's relative invisibility today stems in part from his failure to pursue the publication of his natural history drawings. This failure to publish is surprising, since he began his career in London's book trade, planning a future as a prominent natural history illustrator. When he was a young man, he met George Edwards, a well-known author and illustrator of several natural history books who, on viewing Abbot's drawings, "praised them much" and, according to Abbot, "desired me by all means to continue Drawing, saying no doubt I would be a publisher hereafter of some work on Natural history."[7] Abbot decided that he needed to travel abroad to hone his skills as a zoological illustrator, and after some deliberation about where to go, he decided on the American South as a place that would enable him to support himself as a natural history specimen hunter for London's natural history collectors and dealers. After two years in Virginia that had proven quite disappointing in terms of finding new and rare insects for his clients, he decided to try his luck in Georgia.

After a difficult overland journey on horseback through the Carolinas, Abbot arrived in Augusta in January 1776. He had hoped that the colony of Georgia with its undeveloped forests, grasslands, and swamps would provide him with a natural environment congenial to collecting and drawing insects.[8] Like other specimen hunters and natural history artists who traveled to places such as Surinam, Sierra Leone, and Australia to collect and draw natural history specimens for Europe's scientific elites, Abbot assumed he would be gone from London for a couple of years while he collected enough insects and painted enough pictures of butterflies, beetles, and wasps to satisfy his patrons in London.[9] Abbot, however, did not return to England; he settled in Georgia, remaining there for the duration of his long life, dying in 1840 on a friend's family farm in Bulloch County.

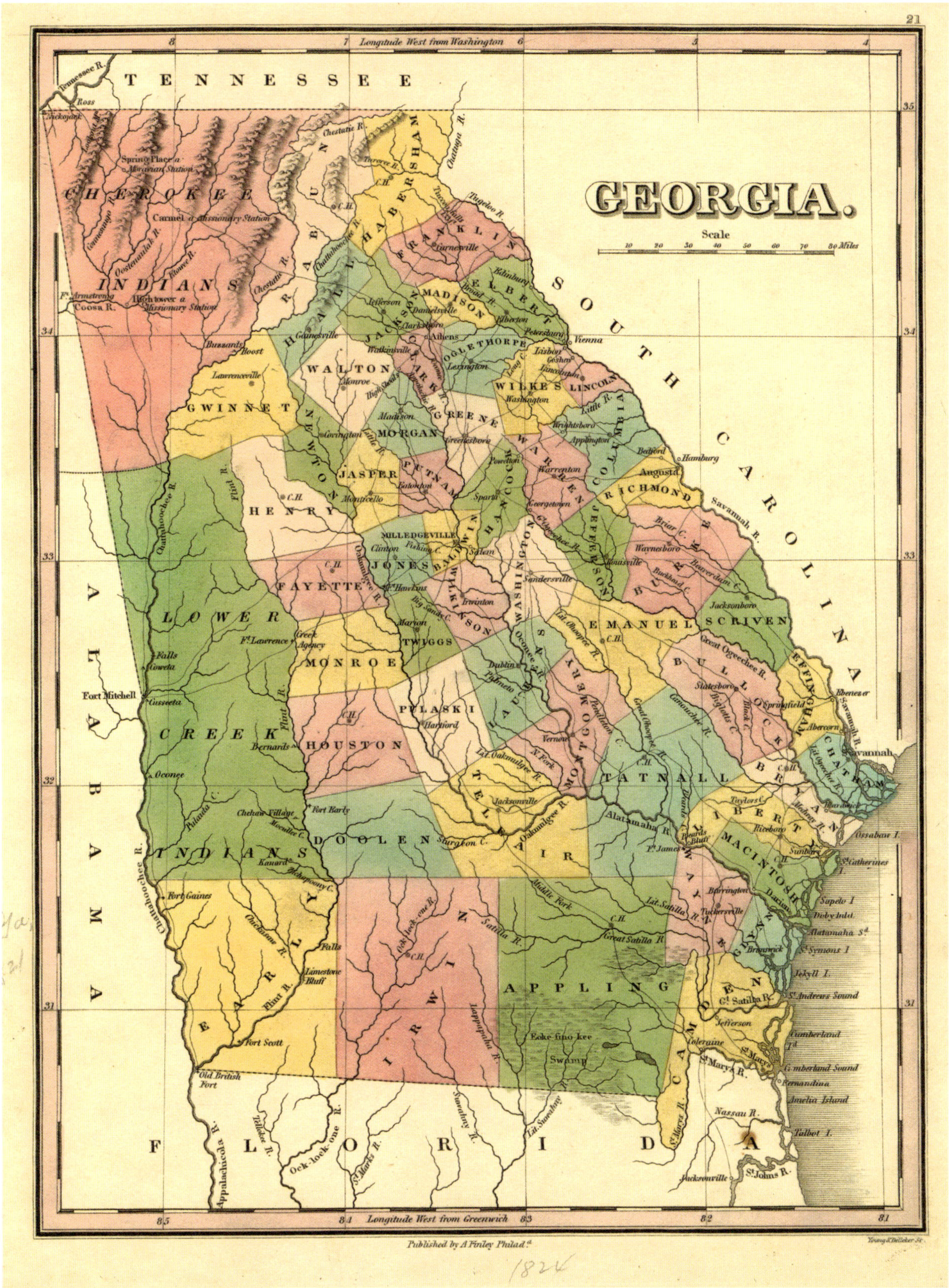

FIGURE 1.2.

Map of Georgia, hand-colored, 28 × 21 cm, Young & Delleker, engravers (Philadelphia: A. Finley, 1824), Special Collections Map 1824 g4, Hargrett Rare Book and Manuscript Library, University of Georgia Libraries. Abbot lived in southeastern Georgia.

FIGURE 1.3. Abbot, "Cicada," northern dusk-singing cicada (*Megatibicen grossus*), det. upheld J. K. Wadleigh, watercolor drawing, "Drawings of the Insects of Georgia, in America, 1792–1804," vol. 5, fol. 46. © The Trustees of the Natural History Museum, London.

Why did Abbot decide to stay in Georgia? Most specimen hunters returned home (unless they died abroad—a professional hazard, as we shall see), and most illustrators were eager to get their drawings engraved, printed, and published in a book format, something that was best done in places like London, with its burgeoning print industry complete with skilled engravers and easy access to fine paper and exquisite watercolors. The American War of Independence (1776–83) interfered with Abbot's plans to return to Britain and prevented him from shipping specimens and drawings back to London. The question remains why he did not return to London after the Revolutionary War or move to Philadelphia, the natural history center of the new republic, decisions that put at risk achieving his youthful goals.[10] With only a handful of documents remaining that can shed light on his private and professional life, we do not have easy access to Abbot's thoughts and feelings. All that survives of his papers are a two-page will, a six-page beginning to an autobiography that he never completed, a clutch of letters that he wrote to naturalists, and the field notes he wrote to accompany his drawings.[11] At the time of his death, Abbot had no family in Georgia (his wife and son had died decades before him) to protect his legacy and to keep his papers safe for posterity, papers that presumably included drawings and templates, field notebooks, breeding journals, and account books with the names of his clients and lists of what he sent them. Without such documents to help us understand Abbot's career choices, we must rely on the corpus of his work as an artist and a naturalist to understand what motivated him to spend the rest of his life in Georgia hunting and drawing insects.

The thousands of drawings and accompanying notes and the few specimens that have survived tell us much about the depth of Abbot's immersion in the natural world and the intensity of his commitment to documenting insect life. To draw as Abbot did required intense concentration, which allowed him to focus on the complexity and beauty of an insect's body. Painstakingly careful in his rendering of whole insect bodies, Abbot produced drawings that honor these insects with his sustained attention and his reverential devotion to detail. His drawings invite us to consider the intricacy and elegance of these nonhuman bodies and appreciate how wondrous they are in their difference. The love for the insect world that emanates from his drawings informed his activities as a naturalist and specimen hunter. Those of us who are not entomologists may regard the catching, killing, and pinning of moths, beetles, crickets, and wasps as a strange way to love insects, but these activities enable entomologists to pay attention to and learn about insects. We can defend Abbot's entomological activities as a form of love in much the same way that Jean-Jacques Rousseau defended his botanical studies: "I am as much a botanist as anyone needs to be who wants to study nature with the sole aim of continually finding new reasons for loving it."[12] We can understand Abbot's decision to remain in Georgia rather than return to London as his choice to live among those creatures he found endlessly fascinating.

Loving Insects

To assert that Abbot chose to remain in Georgia for the whole of his long life because he loved insects may seem a somewhat outlandish proposition. Yet I came to this conclusion when I spent a year shadowing coleopterists (beetle people) and lepidopterists (butterfly and moth folks) at my university as part of my preparation for studying eighteenth- and nineteenth-century entomology. During my research for this book, I showed some entomologists photographs I had taken of Abbot's drawings held in London's Natural History Museum. "There's my favorite moth!" said Dr. John Pickering as he scrolled through my computer's digital photographs of Abbot's drawings. At that moment I caught a glimpse of the passionate engagement entomologists have with their objects of study. I had asked Dr. Pickering, a lepidopterist in the Odum School of Ecology at the University of Georgia (now professor emeritus), for help in determining the scientific value of Abbot's drawings. Pickering, who specializes in moths, is cofounder of Discover Life, an online site designed to help citizen scientists and professional scientists with botanical and zoological identification. As we talked about Abbot's drawings, Dr. Pickering, or Pick as he is called by friends and associates (including his graduate students), explained that the state of Georgia has an extraordinary number of moth species, especially for a temperate climate. Hearing Dr. Pickering, an Englishman, talk with enthusiasm about the number and rarity of moths that gathered around his porch light at night, I realized that Abbot may have felt similarly about Georgia's moth population, and like Dr. Pickering, Abbot chose to live in Georgia to be surrounded by what he loved to study.

Entomologists are passionate about the objects of their research, whether it be moths or wasps or dragonflies, and often they have feelings of admiration and affection for the insects they study. The students in Dr. Joe McHugh's graduate course on insect taxonomy, which he so graciously allowed me to audit, wore T-shirts emblazoned with images of ants, beetles, or termites—whatever was each student's favorite arthropod. Their shared enthusiasm for the study of insects was a revelation for me and a delight to behold as these young women and men scampered about during field trips waving butterfly nets, beating bushes, and prying bark from fallen trees in search of insects. Each student was required to make a collection of insects as a class assignment. I watched the students in the field as they transferred their captured insects into kill jars and then stored the dead insects in sealable plastic bags. Later in the lab students struggled to identify what they had caught, studying their catches under microscopes. In the lab they conferred with their fellow students, enjoying the process of identifying the insects and offering suggestions, encouragement, and sympathy for difficult cases. These qualities of collaboration and enthusiasm that I witnessed in the academic study of insects are not unique to this group of students or even to this particular historical moment but are endemic to the study of insects. Of course, there's competition then and now about the correctness of an identification and the privilege of naming a new species, but in general, entomology is a highly social activity with its own culture, one that began in earnest in mid-eighteenth-century Europe with enthusiastic collectors, some of whom Abbot knew quite well.

"A Rank of Beings So Wonderful"

So that we may understand more fully Abbot's passionate engagement with insects, it is helpful to recover how contemporary insect enthusiasts described why they were so fascinated with them. Dru Drury (1724–1803), a London silversmith, insect collector, and Abbot's mentor, defended his love of insects by arguing that "there is no part of the kingdom of nature, where so many wonders are unfolded to us, as in this of insects."[13] As we shall see in the next chapter,

Drury went to enormous lengths to acquire his insect collection, sponsoring collecting expeditions, supplying travelers with directions and equipment, and courting fellow amateur naturalists worldwide to add to his insect collection, Britain's largest. Abbot was one such traveler whom Drury helped equip, and he was the delighted recipient of thousands of insects that Abbot had collected.[14] Elaborating on the wonderful world of insects, Drury wrote in the preface to his multivolume publication *Illustrations of Natural History, wherein are exhibited upwards of two hundred and forty figures of Exotic Insects* (1770–82), usually referred to as *Exotic Insects*:

> Insects may, with great truth, be considered as a rank of beings so wonderful and extraordinary, as to strike with astonishment every observer, if we regard either their structure, powers, or use; and creatures, who, at the same time that they challenge our regard, loudly proclaim the wisdom, goodness, and omnipotence of their great Creator. If their shape and beauty are capable of attracting our notice, their ways of living are no less adapted for exciting our admiration; and the more we enquire into their nature and history, the more occasion we shall find for confessing this great truth, "nothing is created in vain."[15]

Drury listed some of the many amazing things that insects can do, starting, of course, with metamorphosis, giving examples of the "wondrous manner in which numbers of them pass their lives." Some are "appointed to live, during the greatest part of their lives, within the bodies of other animals." Others, such as damselflies, experience their larval stage in ponds, lakes, streams, and marshes, where "they feed and thrive, till they become inhabitants of the air, when their lives are terminated in a few weeks," perishing from rapid senescence.[16]

This fascination with insects and the "wondrous manner" in which they live their lives was shared by William Bullock (1773–1849), who, in his *Companion to the Liverpool Museum* (1805), tried to stoke his audience's enthusiasm for his insect collection with a lengthy description of insect anatomy, noting its variety and how well it is adapted to specific environments. Describing the variation in insect legs, he links structure with function:

> The structure of the feet of Insects is truly admirable: those that live in water have their feet long, flat, and somewhat hairy. Such as have occasion to burrow into the earth, have legs very broad, sharp edged, and serrated. Those that use them for walking only, have them long: some have sharp hooked claws, and skinny palms, like unto sponge, to assist them to walk on glass and other smooth bodies, as in the Fly, &c. and the Spider has each foot armed with a kind of comb, probably for the purpose of separating the threads that issue from its body, and prevent them from tangling. In the hind legs of those

FIGURE 1.4.
Abbot, "Mason Fly," yellow-legged mud-dauber wasp (*Sceliphron caementarium*) and its prey, black widow spider (*Latrodectus* sp.), det. J. K. Wadleigh, watercolor drawing, *Moths of Georgia* (ca. 1817–20), vol. 2, plate 105, Hargrett Rare Book and Manuscript Library, University of Georgia Libraries.

FIGURE 1.5.
Abbot, "*Scarabaeus Carnifax*," male (possibly "minor"), male, and female, rainbow scarabs (*Phanaeus vindex*), det. J. K. Wadleigh, watercolor drawing (ca. 1785), "Drawings of Insects and Plants," MS Typ 426.2, Houghton Library, Harvard University.

> which occasionally leap, the thigh is very thick and large, and the shank long; of these are the Grasshopper, Crickets, &c.[17]

Bullock's description of arthropod anatomy extends to mouths, jaws, and tongues. He notes that their anatomical structure fits what the insect eats: Long tongues suck up a flower's nectar, and strong jaws bite and grind wood.

William Kirby and William Spence, in their four-volume *Introduction to Entomology* (1816–26), used the language of tools to register their amazement with insect anatomy: "They have their saws, and files, and augers, and gimlets, and knives and lancets, and scissors, and forceps, with many other similar implements; several of which act in more than one capacity, and with a complex and alternate motion to which we have not yet attained in the use of our tools." They urge their readers "to pay closer attention" to insects and "their ways than we have hitherto done, since it is not at all improbable that the result would be many useful hints for the improvement of our arts and manufactures." Insects, Kirby and Spence argue, may be "considered as a set of miniature patterns drawn for our use by a Divine hand."[18] These early entomologists saw the divine in insects with their tiny but perfect bodies, which were well suited for the conditions under which they lived, possessing a kind of functional elegance that spoke to eighteenth-century entomologists of a providentially ordered natural world.[19]

Abbot shared this sense of wonder and deep fascination with the natural world. Most of Abbot's verbal descriptions of the arthropods he drew are brief, written in sentence fragments that describe where he found them and provide details about their food, habitat, and life cycle. But occasionally this taciturn man was overwhelmed with wonder and surprise at what he encountered in the world of arthropods. In an uncharacteristically lengthy description of dirt daubers, he explains how this wasp captures and paralyzes spiders, placing the spiders' immobile living bodies in the wasp's mud-walled nest to serve as food for its hatching larva: "They make oblong cases of clay which they plaister in layers to Roofs, Ceilings, and other convenient places, when finished they lay an Egg inside at the end, then fill it with Spiders and plaisters them up. The Worm of the Sphex or Vespa by the time it eats them all is fullfed and spins a thin case like gold beaters skin in which it changes into a Chrysalis." Struck by the dirt dauber's ingenuity, he marvels: "This is a wonderful property and provision in Nature to provide the worms of the Sphex or Vespa with fresh and proper food as long as they want or are in need of it." These dirt daubers, or mud daubers, as they are also known, "destroy an amazing number of Spiders," and "many of them of very rare Species, such I imagine must live chiefly on the top branches of the loftiest Trees, as I could never other ways meet with them." Abbot regarded the nests of dirt daubers as treasure chests to be broken open with eager anticipation:

FIGURE 1.6.
Abbot, "Raphidia Cornuta," female eastern dobsonfly (*Corydalus cornutus*), det. agreed and updated J. K. Wadleigh, watercolor drawing (ca. 1785), "Drawings of Insects and Plants," MS Typ 426.2, Houghton Library, Harvard University.

"On opening several of these Cases at once affords a most curious and pleasing sight, to see such a large quantity of Spiders at one view of the rarest kinds and of the most beautifull colors."[20] Inspired by this description, I knocked down a mud dauber's nest that was attached high up where my front porch joined my house, and lo and behold, there were spiders in it!

Abbot's notes also contain details about insects' bizarre or surprising behaviors, such as the hairball beetle, *Scarabaeus gibbosus*, which "makes a large ball of Hair in which, it deposits its Egg, burying it in the Earth," or the dung beetle, *S. carnifax*, which "flys in the Day time in search of excrements newly fallen from Man or Beasts. They form balls of it in which they deposit their Eggs, these they roll with their breach backwards into deep holes, which they make under or near the dung." Another beetle, *Carabus americanus*, earned its local name "gunpowder beetle" because "when touched it emits a Liquor like powder or Smoke." Similarly well named is the "stinking grasshopper," which because "it is very slow and inactive . . . would render it an easy prey to its Enemy had not nature provided another means for its defence, when touched or molested it emits a brown liquid from its Joints, or rings of its body of so stinking and offensive nature that Fowls and Birds refuse to eat them."[21] While the grotesque is not exactly the same thing as the wondrous, it still can create a sensation akin to wonder.

The enumeration of entomological wonders that these insect lovers reiterated helped their readers to imagine the surprising complexity of insect biology and behavior. Often overlooked because of their tiny size and their sometimes annoying or disgusting activities, insects did not get the respect these authors thought they deserved. For these enthusiasts, insect bodies possess an elegant fitness. Insect bodies contain tool-like structures, as in hooks for feet, jaws with large pincers, joints that emit odors, and ovipositors that can drill into hosts' bodies to deposit eggs. These anatomical features, however grotesque they may seem to be, are perfectly suited to the insects' environments and the tasks these bodies need to accomplish in order to survive. The fitness that insects possess, as the philosopher Henri Bergson explained, is expressed when "an infinite complexity of detail" is combined with "a marvelous simplicity of function." The insect performs "what it is called upon to do, without difficulty and with a perfection that is wonderful."[22] The intricacy, delicacy, and ephemerality of insect bodies and the amazing things these bodies can do are what drove Abbot and people who shared his curiosity about the insect world—insect collectors, amateur naturalists, and entomologists—to collect and study arthropods.

This enthusiasm for insects was part of the era's larger cultural embrace of natural history as an edifying form of educational entertainment that was adopted by a broad range of people: young and old, men and women, elites and the middling and working classes. The popularity of natural history in late eighteenth- and early nineteenth-century Europe extended to activities ranging from visiting natural history museums to collecting botanical and zoological specimens to drawing flowers and butterflies and reading poetry that celebrated the natural world, including insects.[23] This enthusiasm for natural history nurtured Abbot as a young man, guiding him as he developed into a gifted naturalist artist and a talented specimen hunter.

Cultures of Collecting

Abbot built a business around doing what he loved: collecting insects and drawing them. Georgia's abundant insect populations, combined with its port city of Savannah, with its easy access to major shipping routes to such commercial centers as Liverpool, London, Philadelphia, and New York City, made Georgia an excellent place from which to transport insects, birds, and other natural history specimens to collectors and naturalists on the North Atlantic rim. Although Abbot moved several times within Georgia, he managed to live in or close enough to uncultivated lands, including oak swamps, canebrakes, blackwater streams, and pine forests, where he could collect a wide range of insects while never being more than a few days' ride from Savannah's port. Having access to Savannah meant he could easily send his specimens and drawings in ships that were carrying cargoes of cotton to Liverpool, and he could purchase in Savannah the fine English paper, pencils, brushes, and pigments that were shipped from Britain, sometimes via Philadelphia or Charleston. Other places in the South, Alabama, for instance, had plenty of insects but did not have direct shipping ties with Britain, and Florida was not part of the United States during Abbot's lifetime. Abbot's natural history business was dependent on an overseas infrastructure inherited from Britain's heavy involvement in the Atlantic slave trade.[24] He shipped his specimens from Savannah to Liverpool, a trade route established during Britain's colonial period and the trade in enslaved Africans, and he relied on Savannah's cotton merchants, whose wealth as brokers was ultimately created by enslaved laborers. These brokers acted on Abbot's behalf in the exchange of specimens for payment from British clients.[25]

Abbot, who was a slaveholder, benefited not only from an infrastructure built on slavery and plantation economics but also from the stolen labor of enslaved persons of African descent. The few details about his private life that can be pieced together from legal documents indicate that he gave up "house keeping" in 1818 after his wife died.[26] As part of the process of breaking up his household and dispersing his possessions, five enslaved men (Ned, Hezekiah, George, Lewis, and Cesar) and three enslaved women (Frankey, Sarah, and Betsey) were "given" to his daughter-in-law.[27] To what extent these enslaved men and women helped Abbot with the work of hunting, raising, preparing, or transporting insect and bird specimens we do not know, but we can assume at the very least that their labor freed him to pursue his work as a specimen hunter and a naturalist artist.[28] If they did assist him, they may have performed tasks that required knowledge and skill, such as finding appropriate plant matter to feed caterpillars that were being raised, turning butterflies and moths into specimens, preparing bird study skins, building the special boxes that transported specimens and drawings, and packing and possibly transporting those boxes to Savannah's wharves.[29] Judging from his notes, however, it is clear that Abbot caught thousands of insects himself without the assistance of others, at least not until he was in his eighties, when he purportedly paid local children to catch butterflies for him.[30]

Abbot sent thousands of insect specimens to naturalists in Britain and other parts of Europe. These naturalists not only accumulated specimens from within Europe, building inventories, for instance, of butterflies from particular regions, but also were fascinated by insects from the Americas, Asia, and Africa. They acquired specimens from beyond Europe as the imperial and commercial ambitions of European nations grew in the eighteenth century.[31] For those European naturalists eager to catalog the influx of non-European insects, amassing huge collections was key to the scientific process of bringing order to the diversity of the earth's insect life. Classifying entomological specimens required access to natural history collections so that distinctions and similarities could be gauged through comparing specimens of the same species. This process of identifying and classifying specimens became the central mission of the life sciences by the mid-1700s with the rise of Linnaean systematics. Although today identifying a species may include the use of high-powered microscopes and genetic testing to help with the process of distinguishing one species from another, this work is still dependent on access to large collections.[32] Abbot contributed to this scientific enterprise with his work as an artist and a specimen collector.

Evidence of his contribution to knowledge-making practices within European centers of natural history inquiry can be found in museums around the world. Twenty butterflies and moths that Abbot collected in Georgia are

FIGURE 1.7.
Abbot's specimen of the "Common American Tiger Moth (*Phalaena Phyllira*)" (*Apantesis phyllira*), Linnean Society, London. Author's photo.

in London's Linnean Society; a few notable Abbot specimens are in the Macleay Museum in Sydney, Australia; and many of his insects are in the entomological collections of London's Natural History Museum and Zürich's Federal Institute of Technology, their identity as collected by him lost when they were merged into working collections.[33] Abbot's drawings of insects and birds are scattered up and down the East Coast of the United States, mostly in university libraries and natural history museums, but the largest collection of Abbot's insect drawings, nearly three thousand, is housed in London's Natural History Museum, and more than five hundred spider drawings reside in Paris's Muséum national d'Histoire naturelle. Recently, specimens collected by Abbot and his drawings of butterflies and moths have been discovered in several European museums thanks to the indefatigable research of lepidopterist John Calhoun.[34] The total number of extant drawings is unknown, since many drawings may be in the hands of private individuals, some of whom may know little about the artwork that graces the walls of their home or lies forgotten in their attic. Abbot's output of drawings is an achievement, even for a long life blessed with health, energy, good eyesight, and a steady hand. While in his seventies, Abbot drew and colored insect drawings on sheets of paper bound into books for friends in Georgia and England, and in his eighties, Abbot was still beating the bushes for insects to send up north to a Harvard naturalist.

This Book

I approach Abbot's work as an artist and a naturalist by reconstructing the natural history networks in which his drawings, observations, and specimens circulated. Although Abbot drew birds and collected them as specimens, this book focuses

FIGURE 1.8.
Abbot, "Scolopendra," Florida blue centipede (*Scolopendra viridis*), det. J. K. Wadleigh, watercolor drawing, vol. 15, Francillon Collection, Archives and Library, Trustees of the Natural History Museum, London. Abbot's note: "Taken 22 April under an old Log. It is rare. The smaller ones of this Genus is the oftenest to be met with."

exclusively on his entomological activities and artwork. Recovering the world in which insect bodies were much-desired objects, this book describes insect collecting primarily in Britain and the United States, focusing initially on Dru Drury (1724–1803) and John Francillon (1744–1816), London jewelers and amateur entomologists who possessed huge collections and were Abbot's mentors, agents, and clients. Sir James E. Smith (1759–1828), president of London's Linnean Society, was an important figure in Abbot's life, as he was instrumental in turning Abbot's butterfly and moth drawings and notes into the gorgeous two-volume *Natural History of the Rarer Lepidopterous Insects of Georgia* (1797). William Swainson (1789–1855), a talented British artist, naturalist, and nature writer, tried to follow up Smith's work with two more volumes of Abbot's drawings, but these efforts came to naught. Failing to acquire a position within the natural history division in the British Museum, he subsequently departed for New Zealand.[35] With Drury's and Francillon's deaths in the early part of the nineteenth century and Swainson's preoccupations, Abbot turned from trying to forge new networks in Britain to developing ties with American, French, and Swiss naturalists and natural history collectors. These networks were complex, often fragile entities that would expand and collapse in relation to world events such as wars, with their embargos on international shipping; life-changing events such as illness or financial ruin; and changes within scientific communities involving methods for generating and organizing information about the animal world.

Chapters 2 and 3 locate Abbot's work as a specimen hunter in the context of the cultures of natural history collecting, tracing his relationships with various British, European, and American naturalists and entomologists. These chapters detail the material practices of specimen hunting, in particular the techniques used to catch, kill, and preserve insects, as well as the methods Abbot employed for transporting his specimens, notes, and drawings to his patrons in Britain, Europe, and the American Northeast. In addition to his skills as a specimen collector, which involved tramping through the woods, swamps, and fields of rural Georgia, Abbot was an immensely skillful rearer of butterflies and moths. Hundreds of his butterfly and moth drawings were of those he had raised himself, an achievement unmatched by his peers.

The vast archive of Abbot's entomological drawings is the subject of chapters 4 and 5. My approach to these drawings is to situate them within the late eighteenth- and early nineteenth-century cultures of collecting that dominated natural history inquiry. Organized around the major collectors of Abbot's drawings, among them Drury, Francillon, John Eatton LeConte, Jean Baptiste Boisduval, and Baron Charles Athanase Walckenaer, these chapters trace the ways in which this particular set of insect collectors and professional entomologists used Abbot's drawings. Natural history drawings in general circulated within naturalists' networks as tools to help identify and name new species and as such were integral to the task of making natural history knowledge. Abbot's entomo-

2

++38.

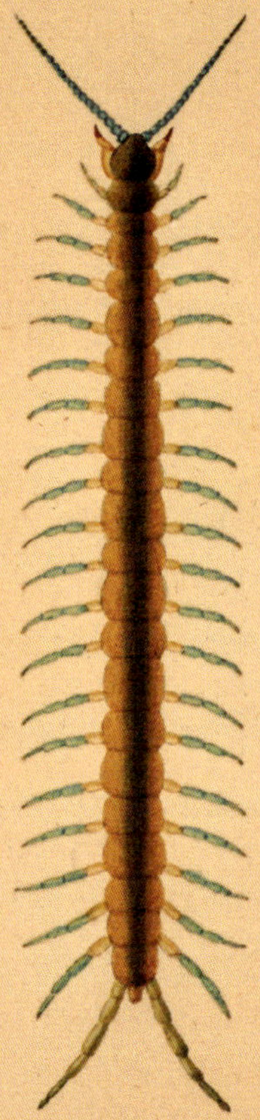

Scolopendra

++38 Scolopendra. Taken 22d April under an old Log, it is Rare, the smaller ones of this Genus, is the oftenest met with.

logical drawings were among the era's most desirable; as Swainson noted, they were "so beautifully chaste and wonderfully correct, that they were coveted by every one."[36] The appeal of Abbot's drawings, then and now, is in the care and attention that he lavished on his subjects, depicting as accurately as possible the complex and sometimes surprisingly beautiful bodies of these tiny creatures.

The science and art of Smith and Abbot's masterpiece, *The Natural History of the Rarer Lepidopterous Insects of Georgia*, known as *Insects of Georgia*, anchor chapter 6's inquiry into how different kinds of expertise were valued (or undervalued) within the emerging field of entomology. Tensions between practical naturalists and taxonomists surfaced in the print culture of early entomology, and these dynamics are at play in *Insects of Georgia*. Although Abbot's observations and drawings were the basis for *Insects of Georgia*, and his observations make up at least half of the text accompanying the engravings of his drawings, Smith is considered this book's author because he added proper Linnaean terminology and named new species. This chapter asks why Abbot was not given credit as the book's coauthor and why his scientific contribution to this volume continues to be underappreciated today. An explanation lies in the competitive world of early entomology, where factions marked by class, gender, race, and regional differences struggled to assert the legitimacy of their scientific practices. This competition for whose knowledge mattered most explains in part the eclipse of Abbot's reputation in the late nineteenth and early twentieth centuries as an important figure in early American natural history. Chapter 7, by focusing on the scientific impact of Abbot's spider drawings, argues that his work as naturalist and artist deserves to be recognized today for its contribution to the study of the North American natural world.

Paying Attention to Arthropods

In addition to reclaiming Abbot's legacy, this book aims to cultivate an appreciation of insects, not just butterflies but also the less popular arthropods such as moths, beetles, wasps, and spiders, a much-maligned and little-appreciated set of beings. Abbot's observations and drawings can teach us to be curious about insect and arachnid life, opening up the natural world of the ordinary backyard and the neighborhood park as sites of potential wonder. Abbot's work as a naturalist and artist has transformed my relationship to arthropods. Before researching and writing this book, I was not a fan of insects, with the usual exception of butterflies and honeybees. While I worked on this book during the pandemic, I sat outside under a muscadine grape arbor in my Athens, Georgia, backyard, where I had many nonhuman visitors. Dropping in on me (literally) were minuscule "inch worms," caterpillars who slid down their silken ropes and dangled over my laptop computer; jumping spiders who hopped about on the table waving at me with their pedipalps (mouth parts that look like small legs); a little yellow jacket wasp who flew off after a quick survey of my lunch plate, disappointed there was no meat to nibble on; a tiny spider who balanced on the rim of my teacup for what seemed a very long time; fuzzy black caterpillars who, thank goodness, I didn't handle, as I found out later that their bristles can cause an allergic reaction; and a solitary wasp who built a nest in the hollow part of my metal chair's back slats. This thread-waisted wasp was carrying tiny twigs as she flew to her destination, using my body as a signpost to help her locate the particular slat in which she was building her nest. I know this because I switched chairs on her when she flew off; when she returned, she went straight to the chair I was sitting on, ignoring the one I had moved aside and in which she had already deposited a few twigs. In her actions I glimpsed how she regarded me, as part of the natural landscape she needed to navigate to achieve her goals. I found these visitors intriguing. What I had either ignored or regarded as nuisances were now sources of interest and pleasure. If the pandemic taught people to love their backyard birds, then it plus Abbot taught me to love or at least appreciate these small creatures. For this gift, I am grateful, and like the newly converted, I am eager to share with others what I have learned about Abbot's insects.[37]

FIGURE 1.9. White-marked tussock moth caterpillar (*Orgyia leucostigma*, Smith 1797) in author's backyard. Author's photo, May 2021.

2

"Going Abroad to Collect Foreign Insects"

LOCAL KNOWLEDGE AND THE GLOBAL CIRCULATION OF INSECT SPECIMENS

A SPIDER RAN across the forest floor trying to dodge Abbot's attempts to capture her. Although this spider ran "nimbly," she did not outrun this man; she was "taken running on the ground 4th Dec." "These kind of Spiders," Abbot remarked, "make no web, run nimbly on the ground & sometimes on the bodies of trees preying on the insects it meets with. They likewise scent their prey on warm evenings and after dark."[1] He found another spider (figure 2.1) on "Dog Fennel 23d Sepr in a cornfield in the oak woods of Burke county with the egg Bags of a dark Brown colur in the shape of a hat, Black cover'd with irregular Knobs."[2] These observations can be found in the "Notes," a handwritten document of sixty-three pages accompanying a bound volume of over five hundred spider drawings that he had prepared in hopes of publishing a book on Georgia's spiders.[3] These notes, containing details about the lives of the spiders he collected, can tell us much about the depth and breadth of Abbot's knowledge of North American arthropods. Abbot's notes are a valuable source of information not only about the lives of these small creatures he collected but also about how he went about collecting the thousands of entomological specimens he sent to Europe's natural history collectors. Using his notes to recover the methods he used to catch insects and spiders, this chapter and the next examine the material culture of insect collecting and trace the global circulation of entomological specimens in the latter half of the eighteenth century and the early nineteenth century.

Abbot's love of insects propelled him out of England and into the woodlands, savannahs, and swamps of the American Southeast. Before leaving London, he had learned already a great deal about insects, in particular how to draw them in a way that would appeal to naturalists (the topic of chapter 4) and how to turn insects into specimens that naturalists would want to purchase for their collections (the topic of chapter 3). This chapter begins with a brief overview of Abbot's London years, when he developed a formative relationship with Dru Drury, who oversaw Abbot's preparation for becoming a practical naturalist and proficient specimen collector. I place Abbot's rela-

FIGURE 2.1.
Abbot, "Spider ++90," green lynx spider (*Peucetia viridans*), det. J. K. Wadleigh, watercolor drawing, "Aranea" (ca. 1799), Hargrett Rare Book and Manuscript Library, University of Georgia Libraries.

tionship with Drury within the larger context of the global exchange of natural history specimens, describing how insect collectors in Britain and Europe developed ties with travelers and sojourners living and working abroad in the hope that they might collect specimens for them. Abbot's success as a supplier of entomological specimens to London's elite natural history collectors was based on skills and techniques he had learned in London, but even more important to this success was his deep knowledge of insects as living creatures and their natural environment, knowledge he acquired by living for decades where he collected, the first forty years in the Savannah River Basin and then the next twenty-five years in the Ogeechee River Basin.

Abbot's Patrons and Clients

When John Abbot went out into the fields, swamps, and forests of Southeast Georgia to search for spiders and whatever insects he could find, he was doing what he loved. He relished the excitement of the hunt for arthropods, scrambling after spiders, running around waving a butterfly net, chipping away a log's bark looking for beetles, and scooping up pond muck where those nippy cousins of water-boatmen lurked. He never outgrew his boyish love of chasing after insects. He claimed toward the end of his life that he always had a "peculiar liking for insects." He reminisced in a very brief autobiography that as a child, "long before I was acquainted with any method of catching or keeping them—I remember knocking down a Libella [a dragonfly] and pin[n]ing it, when I was told it would sting, as bad as a Rattle snake bite." He also remembered how he learned to use a large net from a Mr. Van Dest, "the famous flower painter" whom Abbot bumped into on one of the painter's "Walks after Insects." Van Dest had given Abbot a pattern for making a large net and showed him his own small collection of insects. Abbot recalled these moments when he was an old man, writing that seeing Van Dest's collection gave him the idea that he too could collect insects.[4]

Abbot's love of insects intersected with his "early taste for drawing," which his father encouraged with a "large and valuable collection of prints, of some of the Best Masters" and "many good paintings" and by providing Abbot with a "Drawing Master," Jacob Bonneau, who taught Abbot "the Rules of Drawing and perspective."[5] It was through Bonneau that Abbot made the acquaintance of Dru Drury, a London silversmith and amateur entomologist who possessed "the best Collection of Insects both English and foreign of any one."[6] Drury, according to his nineteenth-century biographer, was "one of the most zealous and successful collectors of insects that ever prosecuted the study in this country." He "rendered most important services to Entomology, and is entitled to the respect and gratitude of all who cultivate the science,—a science which, according to the high authority of Baron Walckenaer, is at once the most difficult, the most extensive, and most philosophical, of all the departments of Natural History."[7] Abbot was delighted to make Drury's acquaintance: "I leave you to judge my pleasure and astonishment at the sight of his cabinets the first I had ever seen of the kind[.] [H]e very politely offered to lend me insects to Draw, and we immediately became well acquainted. That hour may be said to have given a new turn to my future life." Soon after his first meeting with Drury, Abbot acquired "a Mahogony Cabinet made of 26 drawers, covered with sliding tops of glass[.] [I]t cost me 6 guineas & I began to collect with increasing industry. . . . I purchased a parcel of beautiful insects from Surinam . . . and soon began to have a respectable collection. . . . I craved more." Although his father, an attorney, had plans for him to have a career in the law, Abbot decided he had had enough of law clerking: "Deeds, Conveyances & Wills &c was but little to my liking when my thoughts was engrosed by Natural history."[8] With Drury's help, he had at the age of twenty established himself as a professional artist, residing on Poland Street in the heart of an arts community and exhibiting "two drawings of butterflies and insects" as an honorary exhibitor at the Society of Artists of Great Britain's annual exhibition in Charing Cross.[9]

Abbot benefited in multiple ways from his relationship with Drury. In addition to inviting Abbot to draw some of the insects in his massive collection, Drury provided Abbot with access to other naturalists and artists who were in the process of identifying, describing, and drawings insect specimens. Drury's circle included Moses Harris, natural history artist, lepidopterist, author of *The Aurelian: or, Natural History of English Insects; namely Moths and Butterflies together with the Plants on which they Feed*, and illustrator of Drury's three-volume work, *Exotic Insects*; James Lee, horticulturalist, nurseryman, and butterfly collector; Daniel Solander, Linnaeus's pupil, esteemed taxonomist, and curator at the British Museum; and Dr. John Fothergill, Quaker physician and natural history collector who was mostly interested in plants. Visitors to his collections included Margaret Cavendish Bentinck, Duchess of Portland, natural history collector with the largest shell collection of the era, and Johann Christian Fabricius, a Danish entomologist trained by Linnaeus and the author of *Systema Entomologiae* (1775), among other works.[10] Another member of Drury's circle was Henry Smeathman, a young aspiring naturalist whose collecting trip to Africa was sponsored by a consortium organized by Drury, consisting of Fothergill; the Duchess of Portland; Marmaduke Tunstall, a collector of birds; and Sir Joseph Banks, a gentleman botanist whose participation in Captain James Cook's first circumnavigation of the globe catapulted him into the presidency of the Royal Society, a position he would hold for nearly forty years.[11] Smeathman had introduced himself to Abbot as a "brother Flycatcher" (fly as in butterfly). Although Abbot was shy and "not fond of Strangers," he was seduced by Smeathman's charm and enthusiasm: "His Address & discourse soon settled an immediate acquaintance." Smeathman's journey inspired Abbot with the idea of going abroad to collect natural history specimens: "This Mr Smeathman went to Africa on an intention chiefly to collect insects. . . . I now began to entertain thoughts of going abroad to collect foreign insects myself."[12] It took Abbot awhile to figure out where he wanted to go. He realized that he did not relish a long ocean voyage to a place about which he knew little; instead, he opted for Virginia, which, as a long-settled British colony, could offer him some of the comforts he was accustomed to in London. The plan was for Abbot to collect insects and minerals for Drury and other kinds of natural history specimens for George Humphrey and Thomas Martyn, both London natural history dealers, and in this way support himself while he pursued his goals as a natural history illustrator.[13]

Drury helped Abbot prepare for his sojourn as a specimen hunter by providing him with contacts in Virginia.[14] He wrote letters of introduction to Rev. Devereux Jarratt and Dr. James Greenway of Dinwiddie County, men in whom Drury had invested time and money in order to cultivate their talents as suppliers of insects.[15] Drury had hoped that sending them gifts of collecting implements would pay off with several shipments of specimens for him, but he was quite disappointed in the quantity and quality of insects sent to him by these two Virginia correspondents. Writing to Jarratt in 1766, Drury expressed his disappointment with the clergyman's failure to deliver American insects:

> I sent about a month or 6 weeks ago a Box to you containing a fresh set of Implements for catching and preserving of various kinds of Insects. . . . The kind [of] endeavours w[hic]h you have frequently assured me shall be exerted in procuring me the produce of y[ou]r country, when you are furnish'd with proper implements gives me ye [the] most pleasing hopes of receiving in due time some Insects that I never was acquainted with ~~before~~, but as ye Summer is ye time for collecting them I must wait with patience 'till ye return of next Autumn, a circumstance not very agreeable to one whose expectation is so sanguine as mine.[16]

Drury's frustration simmers beneath the polite facade; he was not in a position to criticize these gentlemen who were ultimately doing him a favor. He could not boss them around, although he did tell Abbot that once he arrived in Virginia he should chide Jarratt for not delivering specimens as promised:

> When you see Mr. Jarratt pray tell him I have waited these 2 years with great impatience in hopes of receiving the nest of Drawers with Insects that I sent to him. He gave me some hopes I should receive it well filled but he has not kept his word & you know he could not possibly have filled it according to ye Rules we proscribe without sending many new species. I hope this message will spur him to be diligent. If you have an opportunity, as I have no doubt you will, you can instruct him how to breed & collect the Lepidoptera in *fine* order. This I shall consider as a benefit done to myself & gladly acknowledge it. In short use your utmost persuasion to urge him to return it to me, but not empty.[17]

Drury had hoped that Abbot would teach Jarratt and Greenway how to breed Lepidoptera as a way to increase their productivity as providers of specimens.

Drury's letter to Greenway brims with confidence in Abbot's abilities as an entomologist: "The Bearer Mr Abbot is going to settle in your Country on purpose to collect the various Subjects of Natural History both in the Animal and Vegetable as well as Mineral Kingdoms." He went on to explain that Abbot's "principal Forte" was insects and that Abbot would need Greenway's guidance when it came to botany and ornithology. Although Abbot "may stand in need of your help for collecting the other branches" of natural history, Drury was so impressed with Abbot's knowledge of insects that he told Greenway that Abbot "will be able to improve your knowledge in Insects & thus together you may promote the purpose of Natural History."[18]

However, Drury's grand plans for improving Jarrett's and Greenway's productivity as sources of insect specimens faltered when Abbot, after less than two years, decided to leave Virginia, a place that had proven disappointing in terms of finding new and rare insects for his patrons. Although he was able to send one shipment of insects, which earned him high praise for his industriousness, Abbot had bad luck with two other shipments of specimens: One of the ships carrying his specimens went down off the coast of England, and the other was lost on the James River "when a terrible September storm arose in the night."[19] In addition to these setbacks, the unstable political situation in Virginia worried Abbot. Drury cautioned him to "avoid all disputes &c either in Religion or politics—Tumults & Disturbances arising either from one or the other are never favourable to Natural History."[20] Drury's advice proved prescient when Virginia set a date "after which all intercourse with England, was to be stopped," which would have been the end of Abbot's business as a specimen collector with British clients.[21] It was at this point that Drury encouraged him to go to Surinam, enticing him with the promise of gorgeous tropical butterflies and assuring him that living in Surinam was as commodious as living in New York City.[22] No doubt Drury gave this advice with the hope of having an agent in Surinam who could provide him with more South American insects.

Abbot ignored Drury's suggestion that he go to Surinam. He tried to book passage back to England but failed and then decided to go to Georgia, as it was not yet engaged in revolutionary activity. Traveling with some acquaintances he had made in Virginia who had family in Georgia, he embarked on an overland journey during winter that took two months, arriving in Augusta in February 1776. Abbot's autobiography does not convey much about his years during the American Revolution other than to say: "The first years of my living in Georgia, contains much more of Adventure, than the former part of my life, and continued through such bad & terrible time, that I often reflect upon the goodness of providence, in bringing me safely through them."[23] With the conclusion of the Revolutionary War in the early 1780s, Abbot was able to resume his relationship with Drury after an eight-year hiatus, shipping to him thousands of insect specimens along with drawings and notes. Abbot's decision to go to Georgia eventually paid off, or, as he put it late in life, "To make me amends I have not lost any [cargoes of insects] I have sent from Georgia."

Among those natural history collectors in Britain and Europe who purchased specimens from Abbot was John Francillon. Like Drury, Francillon was a wealthy London silversmith and jeweler who had also amassed a famously huge collection of insects along with other natural history objects. As Drury aged and his activities as a collector wound down in the 1790s, Abbot formed a relationship with Francillon, who became not only a customer of Abbot, buying his drawings and his specimens, but also an agent who acted on his behalf, brokering sales of natural history drawings and specimens within his network of British and European collectors.

Francillon's collection of insect specimens and drawings impressed Charles Peale's sons Rubens and Rembrandt, Philadelphians who in 1802 were in London on a tour of British natural history collections. Rubens wrote home excitedly about the specimens from the "southern states of America":

> A Mr. Francillon has the finest collection in Europe of Insects, from the southern States of America, on Saturday the 6th we began the examination of this collection which consists of Insects from different parts, amounting to 7000 Specimens, classically arranged. The colours were superior to anything I ever beheld, Fifteen folio volumes of drawings of the *worm*, the *crysalis*, and the *fly*, with the Plants which they feed on, besides several hundred not bound. The drawings and a part were exicuted by a [s]Chool Master in Georgia whose name I do not recollect, also the Insects from Georgia &c. This collection I understant he has for sale the price of which is 2000 pounds Sterling or 8,888 Dollars which is moderate.[24]

This Georgia man who executed the drawings and collected insects from the Southern states most assuredly had to be Abbot. Rubens Peale's failure to remember Abbot's name and to misidentify him as a schoolmaster is in keeping with Philadelphia's natural history elite's disregard of Abbot's accomplishments. Not until after Alexander Wilson's Southern birding tour of 1807, when he met Abbot, did a Philadelphia-based naturalist publicly acknowledge Abbot's gifts as an artist and a naturalist. Wilson wondered why no one in Philadelphia seemed to know of his existence.[25]

In addition to his long-standing relationship with Francillon, Abbot established in 1813 a correspondence with Heinrich Escher-Zollikofer (1770–1852), known to Abbot as Henry Escher, a Swiss land speculator who was residing in New York City until returning to Switzerland in 1814. Abbot's friend Augustus G. Oemler (1770–1852), a Savannah apothecary and naturalist, had introduced them, thinking Abbot might benefit from Escher's desire to build a large insect collection.[26] Abbot corresponded with Escher for over twenty years, writing friendly letters and sending him thousands of insect specimens, notes, and drawings, initially to Philadelphia and New York, then to Switzerland when Escher returned to his homeland.[27]

When Francillon died in 1816, Abbot turned to William Swainson as a possible agent to act on his behalf in London's natural history circles. Their exchange of letters from 1816 to 1819 reveals Swainson to be a willing partner, offering to purchase Abbot's drawings with an eye to their publication and expressing a desire to purchase his specimens as well. Swainson, however, was not wealthy like Drury or Francillon and could not afford to buy as many specimens as he wanted:

> I am likewise desirous of possessing all the Coleoptera [beetles] & Phalaena [moths] of Georgia, and my only motive for not wishing you to collect them for me *at present* is, that I do not wish to incur too heavy an expence at one time, when I have procured most of the species of my favourite Genera you can then begin gradually to send me them, the same reason prevents me at present from wishing the drawings of Phalaena, but I trust we shall enter into a contract for these also, when those I have mentioned are finished.[28]

As a way to acquire more specimens from Abbot, Swainson, writing from Brazil, where he had been collecting insects and birds for a couple of years, proposed:

> The extraordinary success I have met with in Entomology during my travels in this Country has suggested to me that could we establish an exchange of Insects of Brazil for those of Georgia it would be to our mutual advantage, for should you not wish a Collection for yourself you might dispose of them very profitably to the public or private Cabinets, I think I could let you have between 4 and five thousand specimens large and smaller and for which in return, I am very willing to take of your Coleoptera Phalaena, or any other insects you may send me, from N° specimens in exchanging Insect for Insect & not more than four of a sort.
>
> If this arrangement suits you pray lose no time in forming an extensive collection of the Coleoptera & the Moths, but I wish you to understand that our original agreement for the Insects I first wrote to you about will still hold good, unless instead, you want rather I should pay you in Brazilian Insects than in money, this is however entirely at your choise.[29]

Abbot directed Swainson to Escher as a possible purchaser of Swainson's Brazilian specimens. Abbot was disappointed to learn from Escher that the two shipments of Swainson's specimens arrived in bad condition, embarrassing Abbot for having introduced Swainson to Escher as a supplier of specimens.

Abbot and Swainson's correspondence seems to have flagged in the 1820s, and their plans for publication never materialized. Swainson admitted in 1840 that the "design" to publish two new volumes of Lepidoptera and their host plants, a sequel to *Insects of Georgia*, "is now abandoned."[30] Having failed to secure a post within the British Museum, he immigrated in 1840 to New Zealand. He took with him more than a hundred insect drawings, mostly of lepidopterans and their host plants, that he had purchased from Abbot in 1818.[31] Today they are housed in the archives of the National Library in Wellington.[32] Along with these drawings, Abbot had sent Swainson "a large box of Insects, containing 900," mostly of Lepidoptera and Coleoptera.[33] Where these specimens went is unknown;

Swainson may have sold them to help fund his journey to New Zealand. Other Abbot specimens, however, did make the journey to the Southern Hemisphere when Alexander Macleay, naturalist turned colonial bureaucrat, moved to Australia in the 1820s.[34] A spectacular red-winged grasshopper collected by Abbot is housed in the Macleay Museum today, Macleay having bought it at the 1818 auction of Francillon's insect collection.[35] More than dozens, if not hundreds, of Abbot specimens that Francillon owned are rumored to be in the Natural History Museum in London, having initially been purchased at the Francillon auction by Adrian Hardy Haworth, George Milne, and Edward Donovan, whose collections ended up in the British Museum.[36]

Abbot also developed relationships with American naturalists, entering into a long-standing correspondence with Major John Eatton LeConte (1784–1860), for whom he collected insects and drew illustrations.[37] In 1816 Abbot described this relationship with LeConte as such: "At present I am under no particular engagement to any person to collect or Draw for, but only to Mr Le Conte a

FIGURE 2.2. Abbot, "Crimson underwing Locust," eastern lubber grasshopper (*Romalea microptera*), center figure is a fifth instar nymph, and the two others show adult variability, det. J. K. Wadleigh, watercolor drawing, "Dessins originaux de plantes et insectes de Géorgie" (ca. 1818), Zentralbibliothek Zürich.

Gentleman who has Plantations in Georgia, but is now at the Northward—he has a Taste in General for Natural history but seems mostly attached to Botany—for him I have Drawed and am continuing to Draw a large Collection of Insects. I believe he has some publications in view."[38] While some of these biographical details supplied by Abbot are not accurate, his having conflated Major LeConte, an engineer in the U.S. Army, with his brother Louis, who inherited a rice plantation near Riceboro, Georgia, Major LeConte did indeed have some publications in view, as we shall see in chapter 4.[39] As late as the 1830s, when he was in his eighties, Abbot wrote to Dr. Thaddeus William Harris, a Harvard librarian, asking if he were interested in setting up an exchange of insects between them, Abbot sending Southern species and Harris sending more northerly species, a proposition to which Harris gladly assented. As we shall see in the next chapter, Abbot's skills as a specimen hunter were extraordinary, since he managed to keep up the supply of specimens for over five decades (1785–1835), supporting himself and his family with his activities as an artist and a specimen hunter, both activities dependent on his expert knowledge of arthropods in Georgia.

Abbot's Notes

Abbot's knowledge of insects and spiders, creatures he found endlessly intriguing, arose out of decades of experience as a specimen hunter. The notes he sent along with his drawings demonstrate the depths of his knowledge of arthropods as living creatures with complex lives. For example, in a note accompanying a drawing of a wasp, now identified as *Ichneumon atracta*, he wrote: "Frequents oak woods and swamps very rare. They breed in stumps of trees that has been cut down. I have taken them when laying their eggs with their tails [ovipositors] thrust into the wood so far that they were sometimes not able to disengage themselves."[40] Of a hornet he caught and drew, he wrote: "Vespa. Taken 14th July. They build in trees & increase the Size of their nests to that of a mans head, of a Substance Like whitish brown paper. They frequent houses to catch the flies for their Young. I have seen them endeavouring to catch the Tobacco Sphinx when sucking the Jamestown weed blossoms in the evening, but though they seized on & brought them to the ground, they were always forced to abandon so large a prey. They are taken all the sum[m]er till the frost and are com[m] on. Called in Savannah baldfaced Hornets."[41] His explanation of how pine sawyer beetles got their name conveys information about their life cycle. These beetles, known today as long-horned beetles, "lay their eggs in the chinks of the bark," and when the larvae eat into the wood, making "irregular furrows," the grubs "may be heard very plain when eating," the noise resembling the sound

FIGURE 2.3.
Abbot, bald-faced hornet (*Dolichovespula maculata*), watercolor drawing (ca. 1785), MS Typ 426.2, Houghton Library, Harvard University.

FIGURE 2.4.
A very large bald-faced hornet's nest (twice the size of a human head) in a camelia tree, Athens, Ga. Author's photo.

of sawing. "When the Larva is nearly full fed, they eat quite into the wood where they change into the Chrysalis."[42] These grubs are "the favorite food of the large woodpeckers, which are very dexterous in barking the Trees in seeking after them."[43]

This kind of information would have pleased Drury, who wrote to a correspondent in Quebec: "Nothing could be more agreeable to me than to know their history, manner of life, what they feed upon, where they live, what enemies they have, how they secure themselves from the power of those enemies, what time of the year they make their appearance, how often in the year they breed, and in short many other circumstances that it is impossible for me to have any idea of, and which only being familiar with the subjects, and seeing them alive, can possibly furnish."[44] Drury's desire for information about insects' lives would have been satisfied with such details as these found in Abbot's notes: The swallowtail's caterpillar "when disturbed" shoots "forth a soft forked horn similar to that of a snail," and a red wasp, the mourning cloak caterpillar's fierce enemy, seizes the caterpillar, "cutting it to pieces to make into a Lump the better to carry to their Nest to feed their young with."[45]

What Drury called "history," narratives about insect behavior, life cycle, and habitat, emerges from Abbot's descriptions of how he caught his specimens. His success as a specimen hunter was dependent on knowing where insects lived

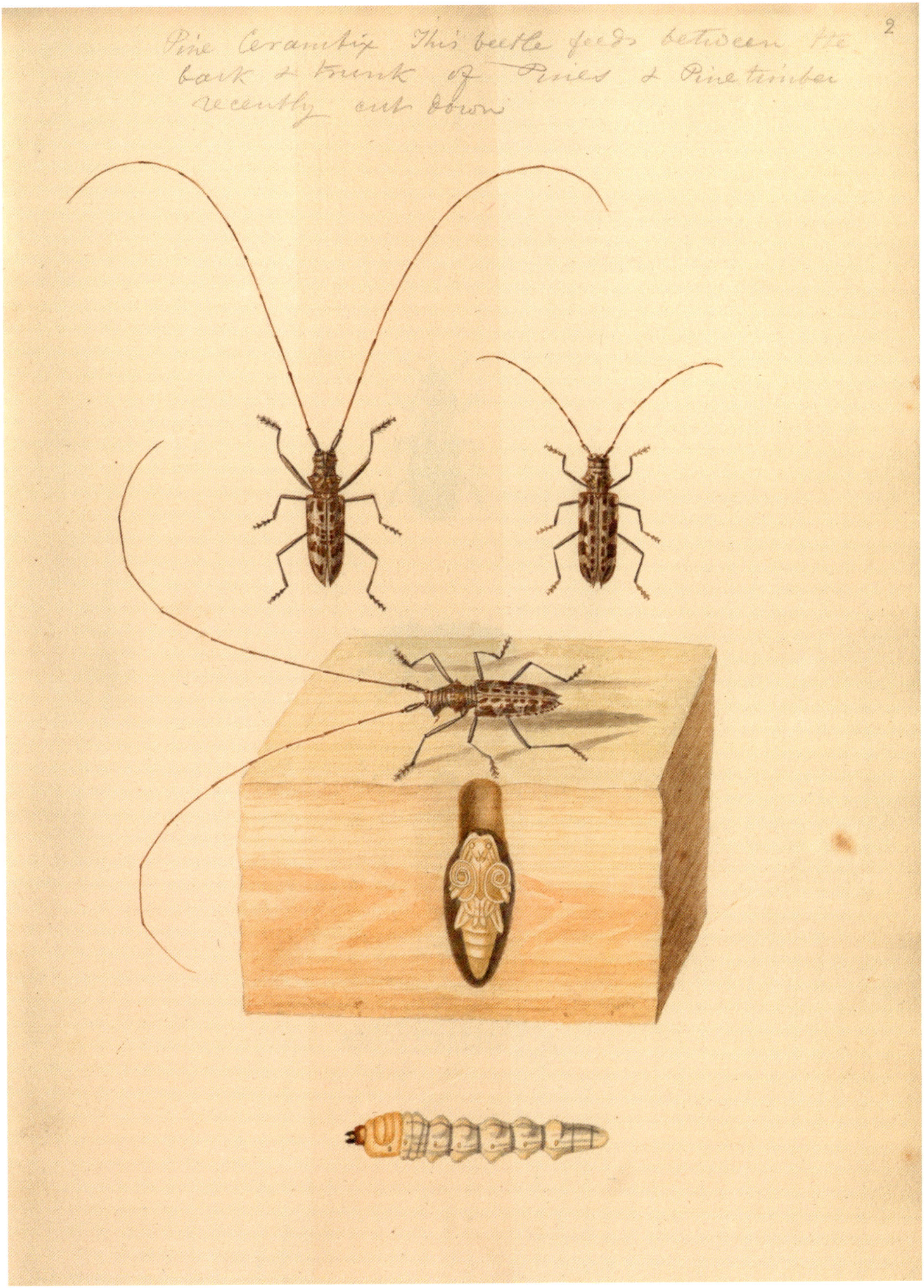

FIGURE 2.5.

Abbot, "Pine Cerambyx" (male, female, male, pupa, larva), southern pine sawyer (*Monochamus titillator*), det. J. K. Wadleigh, watercolor drawing, "Moths of Georgia" (ca. 1817–20), Hargrett Rare Book and Manuscript Library, University of Georgia Libraries.

and how they behaved. In his notes, he indicates where he found each specimen, mentioning, for instance, "Oak Woods in Burke County," "Ogeechee Swamp," "Briar Creek," and "Briar Creek Swamp," along with other details about these creatures' lives. In one of his spider albums, he describes having taken a spider (+229) on December 30 "in its winter retreat between the Bark & Body of an old pine" in the woods of Burke County.[46] He captured another (+233) "running on the ground in a cornfield that had been freshly plowed."[47] In Briar Creek Swamp he caught what he calls "gnat spiders" (+210), who make their webs "about logs & bushes over the water to catch those insects that frequent it."[48] Another spider he collected "makes no Web but hides itself in the blossoms of Flowers, chiefly yellow ones, preying upon the Butterflies and other insects that come to suck the Blossoms. Taken 14th April on a Flower."[49] These notes, while quite concise, written in sentence fragments, provide pertinent details about the arthropods he caught and their manner of living.

Abbot's admiration occasionally erupted in response to the insect behaviors he encountered. As noted in the introduction, he was rapturous over dirt daubers as providers for their offspring with their ability to immobilize spiders and stuff them alive into the wasps' mud-walled nests for their larvae to feed upon. Two other kinds of wasps caught his attention with their elaborate methods for feeding their young. One sphex, a solitary hunting wasp, built its nest "in holes they make in the dirt in the sides of Wells," and "they feed their young entirely with the large sorts of green locusts or grasshoppers, but what is remarkable, with none but green ones," which "are seldom to be met with but where they build."[50] Another solitary wasp, which Abbot identifies as a *Sphex sabulosa*, builds its nest "in the ground, and feeds its young intirely with caterpillars; it is very curious to observe them after having carried into their hole a caterpillar so large that they are hardly able to fly with it, come out carefully & close up the hole by scratching the dirt over so nicely that it cannot be perceived."[51] This species of sand wasp inspired Rev. William Kirby to advocate for the study of wasps, which he argued "are more worthy of our attention" than the rest of Hymenoptera:

> Though they do not, like many of the Coleoptera and Lepidoptera, immediately attract our notice by the brilliancy or gaiety of their colouring, yet when we examine them closely, and observe the consummate skill manifested in their construction; when we attend to their history, replete, be they gregarious or solitary, with entertaining anecdotes, and furnishing instances of the most astonishing sagacity and prudent precaution; we feel inclined to prefer the study of this order of insects to that of any other, not only as most prolific of materials to set forth the praises of Him who hath created them, which is the first duty of the Naturalist—but also as gratifying, in a high degree, our

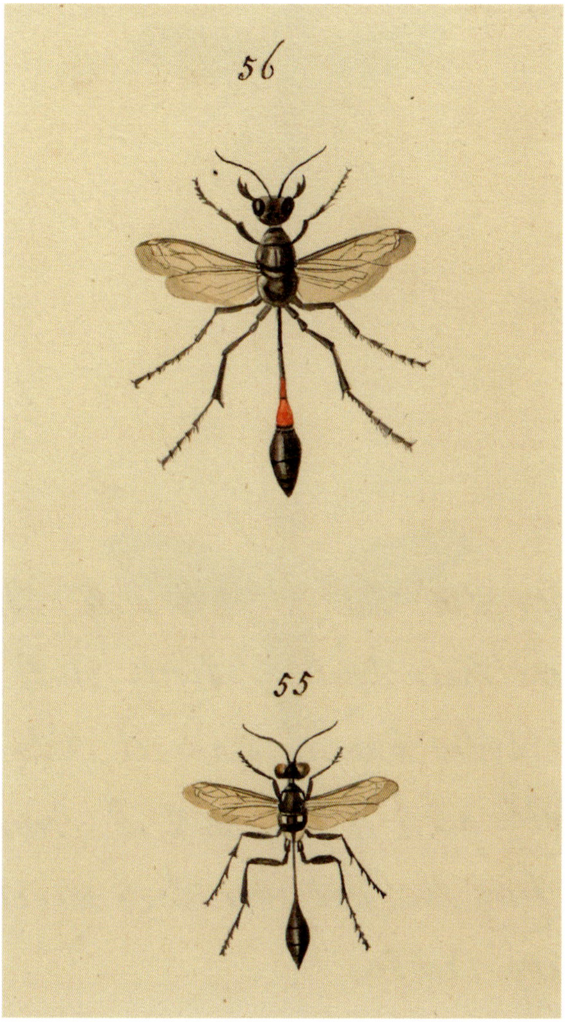

FIGURE 2.6.
Abbot, "Sphex sabulosa," common thread-waisted sand wasp (*Ammophila procera*) (*top*) and gold-marked thread-waisted wasp (*Eremnophila aureonotata*) (*bottom*), det. J. K. Wadleigh, watercolor drawing (ca. 1785), MS Typ 426.2, Houghton Library, Harvard University.

> natural taste for the inspection of things that are remarkable either for their beauty, their structure, or their uses.[52]

Abbot seems to have shared Kirby's enthusiasm for wasps, and he certainly appreciated their ingenuity as nest builders and providers for their young.

Spiders, however, might have been among his favorite arthropods, judging by the affection that suffuses his descriptions. The notes that accompanied his spider drawings, once owned by the Chetham Library in Manchester, England, and now held at the University of Georgia's Hargrett Library, are full of interesting observations.[53] He conveys his admiration for a particularly attractive spider (++68): "None so Pretty Spider. Beat off a Stump in the Oak Woods[.] Beneath the Abdomen white in the middle darker. This Beautiful Spider is the only one I have yet seen of this kind."[54] And there is something poignant in the brief statements made about spider reproduction: Spider ++137 "made its egg bag in the Box I put it in."[55] A ground spider (++14) he caught carried "their young on their abdomen intirely covering it running very fast

FIGURE 2.7.
Abbot, "Spider ++68," black-spotted orb weaver (*Araneus miniatus*), det. J. K. Wadleigh, watercolor drawing, "Aranea" (ca. 1799), Hargrett Rare Book and Manuscript Library, University of Georgia Libraries.

with them."[56] And he described a "Purse Web Spider" that he dug "out of the earth in its web in Decr the body of which was cover'd with a great number of small young ones. The old Spider seemed lively but the Abdomen was much wasted away."[57] Referring to a spider (++25) that he "beat off an Oak 23d Septr in the oak woods of Burke County," he says it was "the only one I have met with but what was very surprizing after I had pinned it to draw 2 slender white worms about an inch in length like those in the intestines of Animals worked out of the side of the abdomen they twisted themselves about a little & soon died. I never met with such an Instance before."[58] These kinds of details about the lives of spiders were dependent on having intimate knowledge of these arthropods, their life cycles, their habitats, and their distribution, details that most European collectors did not have access to, since they usually collected non-European arthropods by proxy.

Abbot's notes describing the arthropods he caught and raised contain details from local lore, ranging from the names children had given to insects and spiders, as in "Father Long Legs" (our daddy longlegs?), to tales told about the poisonous black widow spider:

FIGURE 2.8. Abbot, "Spider 24," female black widow spider (*Latrodectus* sp.), det. J. K. Wadleigh, watercolor drawing, "Aranea" (ca. 1799), Hargrett Rare Book and Manuscript Library, University of Georgia Libraries.

> Makes a Web under old logs & fences, taken 23rd Feb[ruar]y under a stone. [I]t has a large ungulated red spot beneath the Abdomen. This & the following [spider] are accounted very poisonous. I knew a Woman here that lived but a few hours after being bit by a spider, but don't know what species it was. I also was eye Witness to a man in Virginia finding one of this kind in the bottom of a bowl of milk he had eaten from. Fellow was very much frighten'd & vomited up his dinner, but it did not hurt him; the Spider was whole, its not being mashed, & being in milk (I dont doubt) considerably abated its effects even if it had been broken. A Doctor Wells in Augusta I was *creditably* informed gave a girl one of this kind bite by way of experiment in a lingering Disorder, but it killed the Poor Girl.[59]

Moths also figured as villains in these local tales. The caterpillar of the "Royal Persimmon Moth" (regal moth, *Citheronia regalis*), for instance, was called the "Hickory Horned Devil" because on its head were horn-like protuberances, and when disturbed, it assumed a threatening posture. This caterpillar "draws up his head, as in the figure, shakes its head from side to side, & makes a formidable

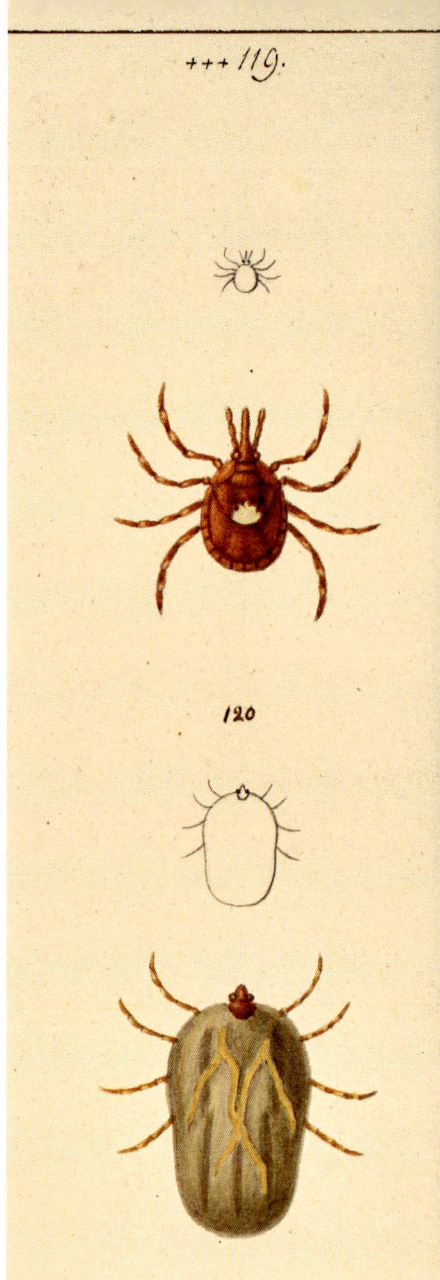

appearance, so much that I never see any one that wou'd venture to handle it. People in general dread them as bad as a Rattlesnake. The horns does not sting at all nor the Worm at all. And when I have handled them before any Negroes to convince them that they wou'd not hurt, they wou'd answer they cou'd not sting me but they wou'd them."[60] Stinging caterpillars loomed large in the local residents' imaginations, even though Abbot tried to alleviate their fears with demonstrations: "I think it a curious matter that many of the Catterpillars here sting like a Nettle, have the effect to raise a white blister on the skin. It seems to sting most when accidentally, or slightly, touched by the bare skin, they therefore are a great terror to many. However, I never yet saw a Catterpillar that I was afraid to handle, often letting the stinging ones fall on the back of my hand, & put them in my bosom, to the great admiration of many both whites & Negroes."[61]

Far worse than stinging caterpillars were ticks, a "great pest of the Country of Georgia":

> They begin to hatch in June, and climb up and cling the whole brood together in large clusters on the Grass and plants in the Woods and Swamps. Upon any person or beast touching them, they will instantly quit their hold, fasten upon them and run over them to find the naked Flesh to feed on. Though they are very small, they bite exceedingly sharpe, and bury their head and Trunk so deep in the Flesh that it is very hard to get them off the parts. . . . They are so exceedingly numerous in some parts of the Oak Woods and Swamps that in a short walk a person will be covered with thousands. . . . The Females now fasten upon Cattle and continue there several days together growing as big as in the Drawing; when full grown they drop off to the ground to deposit their Eggs in a fit place for that purpose. . . . Tobacco smoke will kill them and upon clothes being spread in the hot Sun, they will leave them, or the Sun & heat will kill them.[62]

Mosquitoes, "one of the plagues of Egypt," were, according to Abbot, along with ticks and sandflies, one of "the three great pests of Georgia." He described how "they are excessive numerous in the lower parts of the Country and Swamps of the large Rivers. At times in Savannah River Swamps, if a person stands still a single minute in the Day time He is severly bit in the Neck, Face and Hands." To avoid being bitten in their homes, people living where mosquitoes "are very numerous are obliged to keep a Smoke in their Rooms of an Evening to keep them out."[63]

Some of Abbot's notes convey, however obliquely, a changing landscape and socialscape as white settlers in Georgia moved into the interior from the coast when Creeks, Choctaws, and Cherokees were violently dispossessed of their land.[64] The forested landscape that Abbot first encountered in what was Burke County disappeared, as the lumber industry proved highly profitable for a select few, and, once deforested, the land was harnessed to agriculture.[65] Traces of an altered and disrupted natural environment surface in Abbot's notes and letters: Spiders run across newly plowed fields, and smoke fills the air as tree stumps burn throughout the warmer months. Writing to Escher in 1813, Abbot worried that the annual burning of wooded tracts of land "must destroy an immense quantity of Insects."[66] Abbot grumbled in a letter to Swainson dated November 10, 1818, about the impact that this rapid deforestation had on his ability to find insects and birds in areas that once were wooded and harbored plenty but were now cleared and emptied of wildlife: "I must also observe to You that this Country fails much latterly to what it even furnished formerly, owing I suppose to a great degree to its being cleared and settled, and the woods being burnt

FIGURE 2.9. Abbot, "Great Royal Persimmon Moth, *Phalaena Regia*," regal moth (*Citheronia regalis*), watercolor drawing (ca. 1785), MS Typ 426.2, Houghton Library, Harvard University.

FIGURE 2.10. Abbot, "No. 119 Acarus," lone star ticks (*Amblyomma americanum*) (*top* unfed, *bottom* engorged), det. J. K. Wadleigh, watercolor drawing (1792), vol. 14, fol. 10, Francillon Collection, Archives and Library, Trustees of the Natural History Museum, London.

FIGURE 2.11.
Abbot, "Cotton Caterpillar Moth," cotton leafworm moth (*Alabama argillacea*), watercolor drawing, plate 74, "Moths of Georgia" (ca. 1817–20), Hargrett Rare Book and Manuscript Library, University of Georgia Libraries.

FIGURE 2.12.

Abbot, "Great Brown Emperor Moth, *Phalaena Cecropia*," giant silk moth (*Hyalophora cecropia*), original watercolor drawing for TAB. XLV, Smith and Abbot, *Insects of Georgia* (1797). Courtesy of John Work Garrett Library, Sheridan Libraries, Johns Hopkins University.

every spring for the benefit of their Cattle—It is the same in respect to Birds, I have not made any Collection of them since the peace."[67] He was concerned that intensive agriculture and animal husbandry threatened the biodiversity that he had found so attractive when he first moved to the Savannah River Basin.

Georgia's agricultural landscape was transformed as huge tracts of land were cleared by an enslaved workforce and harnessed to the production of cash crops, primarily cotton and rice. Planters pursued large-scale production of these commodities, choosing monoculture over mixed agriculture, a practice that, while profitable, consumed land, resources, and human labor with devastating consequences. Occasionally, Abbot would comment on insects that threatened this new agricultural regime, describing, for instance, how destructive the larva of the "Ground Worm Moth" was to cotton, corn, and cow peas, "burying itself round the roots in the ground in the day, feeding in the night."[68] The larva of the "Rice Moth" was in some years "so numerous as to do much damage in Rice and Corn fields."[69] To counter the depredations of the "Tobacco Bud Worm Moth," "a Great Enemy to the Tobacco Planters in Virginia, & other parts, eating the bud or main shoot," farmworkers threw wood ashes at the caterpillars, and if that did not work, they were picked "off by hand."[70] Cotton was also attacked by the larva of the "Cotton Caterpillar Moth," which "feeds on the Sea Island Cotton, spun up in a leaf 24th Augt. Bred Sepr. This species was so extremely numerous in the year 1804 as to destroy whole fields of cotton in the course of 10 or 12 days, and extended into the inland parts."[71] We now know that intensive planting of one kind of crop puts these plants at risk of insect infestation, whereas mixed farming with alternating types of crops discourages the march of insects across fields.[72] Despite the threat that these caterpillars posed to commodity-driven, large-scale agriculture, Abbot raised these moths from caterpillars, and he must have fed them their respective favorite foods: corn, tobacco, and cotton.[73]

The failed attempt to bring silk manufacture to Georgia haunts the following passage about a native moth that some had hoped would be as productive as the Asiatic silkworm, the *Bombyx mori*, whose caterpillar spins silk for its cocoon. Accompanying Abbot's drawing of the "Great Brown Emperor Moth" and its caterpillar, shown feeding on a wild American plum (*Prunus pensylvanica*) and identified as *Phalaena cecropia* (today's *Hyalophora cecropia*), are these observations found in the manuscript of *Insects of Georgia*: "Feeds upon the wild plum, tame plum, red root &c. Spun up 17th of June, bred 30th of March following. This caterpillar is hard to raise in the house, as it will not feed freely if confined. Spins up a twig. The outside web coarse, the inner web covered with a kind of silk like a silkworm's Cocoon. I have been informed this silk has been carded and spun into Stockings, & that it will wash like linen. However, this Species is too rare to be of any utility in this Country."[74] The challenges involved in domesticating this elusive North American moth and its rarity banished any hopes that it might have been a substitute for the *Bombyx mori*. In the mid-1700s, colonists had invested heavily in planting mulberry trees with the hope that the imported silk-producing moths from Europe would thrive in Georgia, but these plans also failed to materialize.[75]

Abbot's information on the economic impact of insects was certainly more specific and detailed than European collectors such as Drury expected from their agents stationed abroad. With his skills as an artist and a careful recorder of insect behavior, Abbot was able to share his knowledge of North American insects and spiders as living beings with his European clients, which was quite unusual, since most descriptions of North American insects by Europeans were based on collections of dead specimens.

Local Knowledge and Global Networks

Drury defended collecting as a way to study insect biology, behavior, and ecology, as well as insects' interactions with humans. However, he published very little of that kind of information in his three volumes of *Exotic Insects*.[76] These volumes were filled with illustrations and a text dominated by descriptions of insect morphology, all designed to help readers identify their cabinet specimens. To be fair, Drury acknowledged that he rarely had access to information about his insects as living creatures because the majority of his specimens were gathered by agents who knew little about the specimens they had sold to Drury. Over the course of three decades, Drury persuaded scores of colonial bureaucrats and seafaring men, including ship captains, chief mates, naval physicians, and ship's carpenters, to collect insects for him. He gave butterfly nets to hundreds of these travelers, hoping they would send him exotic insects in exchange for payment. According to Drury's nineteenth-century biographer,

> It was Drury's practice, whenever he became acquainted with any one about to visit a foreign country and willing to take some trouble in collecting insects, to supply him with a series of boxes . . . , forceps, nets, &c. as well as written instructions how to proceed; promising at the same time to defray all the expense that might be incurred, and reward the collector in a manner proportionate to the value of his acquisitions. Such an individual was generally empowered to engage others, after reaching the place of his destination, on similar terms.[77]

Among Drury's papers is a record of these kinds of transactions. "Copy of a List of Boxes delivered on board different Ships since September 1768" lists the names of the men he contacted, their ships and destinations, and the equipment he gave to them. Over the course of a year, he contacted men traveling on twenty different ships sailing for places such as Jamaica, Honduras, Bengal, Madras, Senegal, Sierra Leone, China, and the Gold Coast of Africa. On December 23, 1768, for instance, he gave a special net to Captain Williams, whose ship, the *Meredith*, was sailing to Africa, and on February 8, 1769, he gave three nets to Mr. Richardson, chief mate of the *Conway*, bound for the Bay of Honduras.[78] Ships like these were engaged in the slave trade, military incursions, and other imperial ventures.[79]

To help his agents find, capture, and preserve insect specimens, Drury wrote and printed up a leaflet, *Directions for Collecting Insects in Foreign Countries*, to distribute to potential insect collectors. Drury's instructions covered how to find insects, how to capture and kill them, and how to store them for shipment. Insects are to be found, the instructions say, "sometimes frequenting flowers," while "others, particularly beetles, are found in the dung of animals, and in stinking flesh or putrid carcasses of dead creatures." Some hide under the bark of trees and in rotten wood, and others can be "found under moss that grows on the ground, or at the roots of trees." Butterflies and moths caught in a net should be killed instantly by giving "the body a squeeze with his finger and thumb while it is in the net; which will prevent its fluttering and will kill it." However, other insects such as beetles should not be squeezed; instead, they should be taken home and killed either by holding them over a fire ("but not so close as to scorch or burn them") or "by throwing them into boiling hot water[.] [T]hey will be killed in a few moments, without injuring them, if they are taken out as soon as they are dead." The dead insects should all be pinned in a large square box while they dry. It is important, the leaflet says, to make sure the insects "are quite dead" before putting them in the box, because if they are alive, "they will pull and tear one another to pieces with their claws." Using these techniques, readers of these directions would be able, Drury contended, to catch desirable insects: "butterflies, moths, beetles, flies with transparent wings, locusts or grasshoppers, wasps, ants, fireflies, or in short any other sort except *Cockroaches*, *Centipedes*, or *Scorpions*," which he did not want.[80]

Drury was not alone in his attempts to educate potential specimen hunters with printed instructions on collecting insects and other natural history specimens. The first printed instructions for collecting insects in English were published in 1741–42 by Benjamin Wilkes, author of *One Hundred and Twenty Copper-Plates of English Moths and Butterflies* (ca. 1773), followed by those in the introduction to Moses Harris's *The Aurelian* (1766). In the 1770s, natural history dealer George Humphrey and collector Ashton Lever issued their own sets of instructions for collecting all kinds of natural history specimens, while John Coakley Lettsom, a friend of Fothergill's, published *The Naturalist's and Traveller's Companion, Containing Instructions for Collecting and Preserving Objects of Natural History* (1772, 1774), which included twelve pages on insects. In 1794

Edward Donovan, collector, artist, and author, published a pamphlet on insect collecting that was read widely by novices and accomplished entomologists alike. The methods described in these manuals involved waving nets in the air, beating bushes, digging for pupae, and pinning specimens, material practices (described in greater detail in the next chapter) that are still in use today, having undergone surprisingly little change in 250 years.

As important as equipping these travelers for insect hunting was to Drury's goals of building the largest insect collection of his era, these methods did not always pay off. Drury complained: "There are very few, even of those who collect abroad, who will give themselves the trouble of making observations; and such as do not want [lack] inclination or ability, and even travel expressly for the purpose, find great, and to us almost inconceivable impediments."[81] For instance, when Captain Mayle of the *Hound*, a sloop of war with a destination of Africa, sent

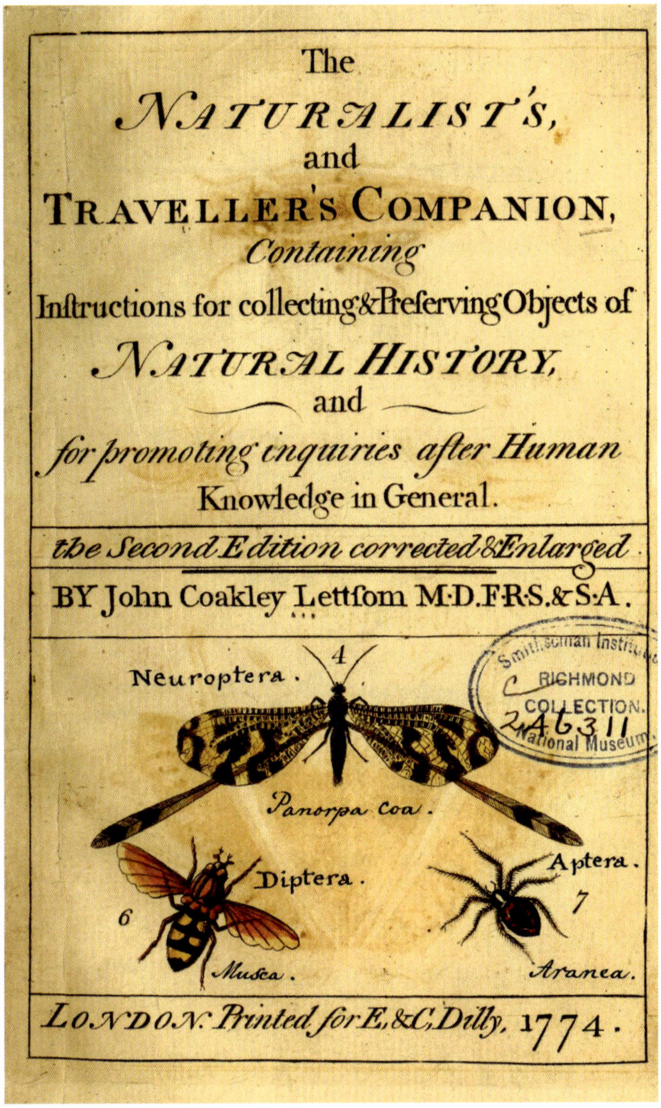
The NATURALIST'S, and TRAVELLER'S COMPANION, Containing Inſtructions for collecting & Preſerving Objects of NATURAL HISTORY, and for promoting enquiries after Human Knowledge in General.

the Second Edition corrected & Enlarged

BY John Coakley Lettſom M.D.F.R.S.&S.A.

Neuroptera. 4 Panorpa Coa.

Diptera. 6 Musca.

Aptera. 7 Aranea.

LONDON: Printed for E. & C. Dilly, 1774.

FIGURE 2.13. Title page, John Coakley Lettsom, *The Naturalist's and Traveller's Companion* (London, 1774), Smithsonian Libraries and Archives.

Dru Drury a box of insects, the naval officer could not say where they were actually found, what their habitats were like, what they ate, or how they reproduced.[82] As a result, Drury's *Exotic Insects* lacked information about living insects. "In this respect the work was necessarily very meager," concluded Sir William Jardine, Drury's biographer, who realized that most of Drury's agents "were contented with transmitting specimens of the perfect [adult] insects without furnishing any account of their previous states [as larvae] or general habits."[83] Jardine argued that Drury had wished to know more about his insects as living creatures, but the way in which the majority of these insects entered his collection prevented the acquisition of this kind of knowledge.

This scattershot approach left Drury feeling the need for agents who resided in the region where they collected and over whom he could have more direct control. His sponsorship of Smeathman's journey to Africa in the 1770s, John William Lewin's to Australia in the 1790s, and Abbot's to the American South was his solution to this problem. He could tell these agents what specimens he wanted, scold them when they failed to deliver, and advise them on how to improve their techniques.[84] Drury's correspondence with Smeathman and Lewin is littered with complaints and correctives, whereas Abbot emerges as someone whom Drury could trust to deliver perfect specimens accompanied by notes describing the natural history of insects as living creatures. Had Abbot's notes reached Drury in time for inclusion in the third volume of *Exotic Insects*, they might have provided much of what the previous volumes' descriptions lacked: textual descriptions of North American insects enriched by details about their "general habits" and manner of living. However, Abbot's illustrations, notes, and specimens did not find their way across the Atlantic to Drury during the American Revolution, and for this reason, they did not make an appearance in Drury's third volume, published in 1782.[85]

This lack of knowledge about exotic insects as living creatures was endemic to the outsourcing of the actual work of finding and capturing them. Most London-based natural history collectors such as Drury relied on travelers to collect non-European insects for them. This practice of calling upon seafaring men and colonial agents had been systematically pursued by natural history collectors in the late seventeenth century. John Woodward (1665–1728), professor of physic at Gresham College and fellow of the Royal Society, and James Petiver (1665–1718), a London apothecary and fellow of the Royal Society, urged travelers, sailors, imperial bureaucrats, and colonists to collect natural history specimens. Recognizing that these agents would not necessarily have the time to wander about in the woods or along the seashore collecting natural history specimens, Woodward suggested in his instructions to travelers that "gathering and preserving . . . may be done by the hands of servants," since not "one single person will have leisure to attend to so many things." Petiver urged his agents to teach enslaved persons "how to collect things" such as shells, insects, and plants.[86] He suggested to Mrs. Carter of Jamaica that she send "any Servant or Negro on a Sunday or at any time which you can spare him once every week or Fortnight to goe into the neighboring woods for an hour or two &... fill a Book with 2 or 3 Spriggs of a sort of whatever Trees Bushes or Herb he shall see."[87] This was a common recommendation made by metropolitan collectors to their agents in the field, Drury having done just this when he told Lewin, his agent in Australia, to hire a man to hunt for insects on his behalf. Lewin replied that "paying a man" to find and gather caterpillars and chrysalids for him was not as simple as Drury had suggested. "For me it is not so easy a matter as you may think," referring perhaps to the labor shortage in Australia (a penal colony with indentured labor practices) or Lewin's own straitened pecuniary circumstances.[88]

Abbot did his own collecting, which was unusual. In general, the actual work of finding and gathering North American insects for European collectors had been done most likely by Indigenous people and by white settlers' children, their servants, and their indentured and enslaved agricultural workers, who were enlisted by travelers and sojourners to perform what they could not.[89] Local people possessed cultural, botanical, and zoological knowledge about the specimens they gathered, but this knowledge was rarely transferred to "centers of calculation" by those tasked with shipping specimens to Europe's capital cities.[90] Specimens arrived in London shorn of all information about them other than the country or region of their origin, designations so broad—"Africa," "China," or "Brazil"—as to be of little use to naturalists. George Samouelle (1790–1846), curator at the British Museum, explained in his collecting instructions (1826) that this lack of local knowledge about exotic specimens stemmed from the brevity of the agent's stay in the place of collection: "The entomological specimens received from *America* have been chiefly contributed by gentlemen who have resided but a short time in the country, and whose excursions in search of insects were, probably, but few and limited; we cannot, therefore, expect yet to possess any thing like a knowledge of American Entomology; yet we have enough to give a stimulus to further researches, and to afford us abundant evidence of the treasures to be found in those countries of the New World not yet visited by the Entomologist." Samouelle seems to have been unaware of Abbot's contribution to American entomology, but in general, he was right that Europe's museums and private collections were filled with North American insects about which very little was known by their possessors.[91]

This shearing off of information was part of the process by which living creatures with their own complex natural and cultural histories were turned into specimens, commodities, and "transportable objects."[92] The labor that locals expended in collecting specimens and the knowledge their actions required usually went unacknowledged by metropolitan collectors and dealers and their agents.[93] For example, Samouelle's printed instructions on collecting insects are "rendered as explicit as the nature of the subject would permit" so that "gentlemen may be enabled to instruct their servants, and especially the *natives* of the country where they reside, to collect the Insects and Crustacea which we are about to enumerate."[94] Implicit in this passage is a long-standing assumption within metropolitan natural history circles that Indigenous people, working-class Britons, and enslaved Africans and their descendants were insufficiently knowledgeable to contribute to the construction of natural knowledge in any meaningful way other than as laboring bodies. The failure to record and transmit to Europe's metropoles the botanical and zoological knowledge possessed by Indigenous and enslaved persons in effect erased the contributions of these "others" to the Enlightenment project of building a storehouse of natural knowledge. This failure is linked to the inability of those in the metropoles to acknowledge the extent to which their scientific endeavors were dependent on long-distance networks that included native, enslaved, and indentured people, soldiers, sailors, dealers, and artists, who formed a chain along which natural knowledge traveled or failed to travel. The scientific knowledge that emerged in the museums and laboratories of London and Paris was dependent on these networks that brought things from far off to up close to be examined, identified, and named within Europe's centers of natural history inquiry.[95]

Collecting zoological specimens required knowledge about where, when, and how to find, capture, kill, preserve, and transport the sought-after creatures. Abbot possessed an intimate knowledge of the natural environment of a relatively small region that centered on the Savannah and Ogeechee Rivers, a region once known for outstanding biodiversity and a place he had spent decades traversing in search of natural history specimens. Because Abbot was knowledgeable about the lives of the creatures he collected, he was the most successful supplier of North American specimens in his era. The next chapter delves into the processes by which living insects were turned into scientific specimens, processes that Abbot had mastered so completely that his skills as a taxidermist were legendary.

3

“The Harvest of Many Years of Industry”

THE MATERIAL PRACTICES OF INSECT COLLECTING

ON ONE OF MY MANY VISITS to London's Linnean Society, Isabelle Charmantier, head of collections, guided me downstairs into a well-guarded room that contains Linnaeus's books and specimens along with a few special items that were in founder Sir James Edward Smith's natural history collections. I was there to look at the moths and butterflies that Abbot had collected in Georgia and that were in Smith's collection. According to the Linnean Society's website, there are eleven lepidopterous specimens with tags in Smith's handwriting identifying the specimens as having been collected by Abbot or as originating in Georgia (which is almost as good as saying they were collected by Abbot).[1] In one of the several glass-topped drawers holding these precious insects is one of my favorite moths: the hummingbird clearwing moth (*Hemaris thysbe*). The first time I encountered a live hummingbird clearwing was in Home Depot's garden section in Athens, Georgia. It had caught my attention as it hovered like a tiny hummingbird over a potted plant's blossoms. Abbot, too, was impressed by its rapid flight: "This insect flies very swiftly in the day time, sucking honey from flowers, and laying its eggs." He called it a "Humble-Bee Hawk-Moth," suggesting he thought it looked more like a bumblebee than a hummingbird.[2] I think his drawing captures the charm of these clearwing moths.

FIGURE 3.1. *Hemaris thysbe*, hummingbird clearwing moth collected by Abbot. Linnean Society, London. Author's photo.

FIGURE 3.2.
Abbot, "Humble-Bee Hawk-Moth (*Sphinx Fuciformis*)," snowberry clearwing moth (*Hemaris diffinis*), det. J. K. Wadleigh, original watercolor drawing, TAB. XLIII, Smith and Abbot, *Insects of Georgia* (1797). Courtesy of John Work Garrett Library, Sheridan Libraries, Johns Hopkins University.

The moths and butterflies collected by Abbot and held at the Linnean Society are a remnant of a massive undertaking involving thousands of moths and butterflies that Abbot had raised, killed, preserved, and shipped up north and across the Atlantic. He raised every moth and butterfly depicted in *Insects of Georgia*, noting, for instance, that he caught the caterpillar of the hummingbird clearwing moth on an eastern bluestar (*Amsonia tabernaemontana*) in May. Shortly after he caught it, the caterpillar "spun itself up on the surface of the ground," with the moth emerging on June 7.[3] Moreover, he raised the majority of the lepidopteran specimens he sold to clients, as he explained to a prospective buyer: "The Lepidoptera are mostly bred specimens in good and perfect Condition. . . . I think if the boxes are carefully opened, I have cemented, pasted, and packed them up so carefully; you will certainly receive them in good Condition."[4]

Not only are the specimens in the Linnean Society a product of Abbot's expertise as a breeder of butterflies and moths, but their existence today is a testimonial to Abbot's skills as a taxidermist, skills his contemporaries were impressed by and grateful for, as Swainson notes in his volume on taxidermy: "Abbot's specimens were certainly the finest that have ever been transmitted as articles of commerce to this country; they were always sent home expanded, even the most minute; and he was so watchful and indefatigable in his researches, that he contrived to breed nearly the whole of the Lepidoptera. His general price for a boxful was sixpence each specimen; which was certainly not too much, considering the beauty and high perfection of all the individuals."[5] Abbot was an exceptionally skilled taxidermist. The moths and butterflies in the Linnean Society's collections, the red-winged grasshopper in the Macleay Museum, and the bird skins in the National Museums Liverpool testify to his ability to preserve zoological specimens that have survived over two hundred years.[6] He even managed to taxidermize large spiders: "The Abdomens of the large thick skinned Spiders I stuff with Cotton—they preserve so very well."[7] Preserving large spiders by removing their internal organs and stuffing the cavity with cotton required finesse, since most spiders were too delicate and too small to taxidermize and were usually dropped into a vial filled with alcohol to preserve them. His taxidermied caterpillars were legendary, as indicated by William Kirby and William Spence, authors of the well-regarded *Introduction to Entomology* (1816–26), who praised "the ingenious Mr. Abbott of Georgia" for his "excellent method of preserving caterpillars, so that his specimens retain their colours and other attributes, and look as if they are alive."[8]

Abbot's success as a supplier of exotic insect specimens was dependent on his having learned in London a set of skills that enabled him to find, capture, raise, kill, and preserve arthropods.[9] He acquired these competences through the informal learning that comes from participating in activities that turned living creatures into scientific specimens and collectible objects. Activities such as "walking after insects" with the flower painter Van Dest, who taught Abbot how to make a large net, were formative for him, as were learning to keep a breeding journal, following Moses Harris's example, and swapping chemical recipes with Drury for ways to preserve specimens from insect infestation. This chapter takes us into this world of insect collecting, detailing the methods Abbot used to turn thousands of living insects and spiders into highly desirable commodities and objects of scientific scrutiny.[10]

Breeding Insects

Abbot was quite an expert at raising butterflies, moths, and a few species of beetles. As we have seen, Drury expected Abbot to teach Reverend Jarratt how to breed butterflies, and this was when Abbot had just arrived in Virginia as a young man in his early twenties.[11] By the time he was in his sixties, he was famous in Britain for the perfect specimens that he had raised. Swainson celebrated Abbot's skills as an insect breeder: "The breeding of exotic insects, more especially *Lepidoptera*, has been prosecuted to a great extent, and with important advantages to science, by a few zealous entomologists: among these, Mr. Abbott in Georgia, North America, and Dr. Horsfield in Java, deserve particular praise. Few can hope to do what they have done in this department, without residing for years in one locality."[12] Abbot had raised hundreds of butterflies and moths by gathering caterpillars in the wild, digging up buried pupae, and finding eggs laid on the leaves of host plants.[13] For example, he discovered the rare "Swallow-tailed Skipper Butterfly" when he saw "the Butterfly lay some Eggs upon the Wild Pea Vine. . . . I afterwards discovered some of the Catterpillers on the species of the Kidney beans."[14]

Rearing insects was ultimately more efficient than prancing through woods and meadows waving a butterfly net in hopes of capturing a swallowtail or a skipper or beating bushes to knock down hapless beetles. Raising lepidopterans from eggs and larvae also produced finer specimens and avoided the problems that could occur when butterflies and moths were hunted with nets that could tear and tatter their wings.[15] Caterpillars, no doubt, would have occupied much of Abbot's attention: finding them, feeding them, and hoping they would emerge unscathed from their pupal stage to become breeding adults and eventually valuable specimens that he sold to clients in Europe and the northern United States.

To understand what it meant for Abbot to have mastered various techniques for raising Lepidoptera, we can turn to natural history manuals from the period for details about this process, manuals such as John Coakley Lettsom's *The Naturalist's and Traveller's Companion, Containing Instructions for Collecting and*

FIGURE 3.3.
Abbot, "Black Scarabaeus or Black Bess" (adult, adult freshly emerged, pupa, larva), horned passalus beetle (*Odontotaenius disjunctus*, formerly *Popilius disjunctus*), det. J. K. Wadleigh, watercolor drawing (ca. 1817–20), Hargrett Rare Book and Manuscript Library, University of Georgia Libraries.

FIGURE 3.4.
Abbot, "Swallow-tailed Skipper Butterfly, *Papilio Proteus*," long-tailed skipper (*Urbanus proteus*), original watercolor drawing, TAB. XVIII, Smith and Abbot, *Insects of Georgia* (1797). Courtesy of John Work Garrett Library, Sheridan Libraries, Johns Hopkins University.

Preserving Objects of Natural History (1772, 1774), Edward Donovan's *Instructions for Collecting and Preserving Various Subjects of Natural History* (1794, 1805), and George Samouelle's *General Directions for Collecting and Preserving Exotic Insects and Crustacea; Designed for the Use of Residents in Foreign Countries, Travellers, and Gentlemen Going Abroad* (1826), as well as books specializing in Lepidoptera.[16] For instance, Moses Harris's *The Aurelian: or, Natural History of English Insects; namely Moths and Butterflies together with the Plants on which they Feed* (1766, 1778) describes how to use various kinds of nets to catch butterflies and moths, explaining in great detail how to use a bat-folding net:

> On seeing the Insect come flying toward you, you must endeavour to meet it, or lay yourself in its Way, so that it may come rather to the right Side of you,

as if you intended to let it pass; then having the Net in your Hands, incline it down to your right Side, turning yourself a little about to the Right, ready for the Stroke; not unlike the Attitude in which a Batman in the Game of Cricket stands, when he is ready to strike the Ball, only his Bat is lifted up, but your Nets must include rather downward: When the Fly is within your Reach, strike at it forcibly, receiving the Fly in the Middle of your Net.[17]

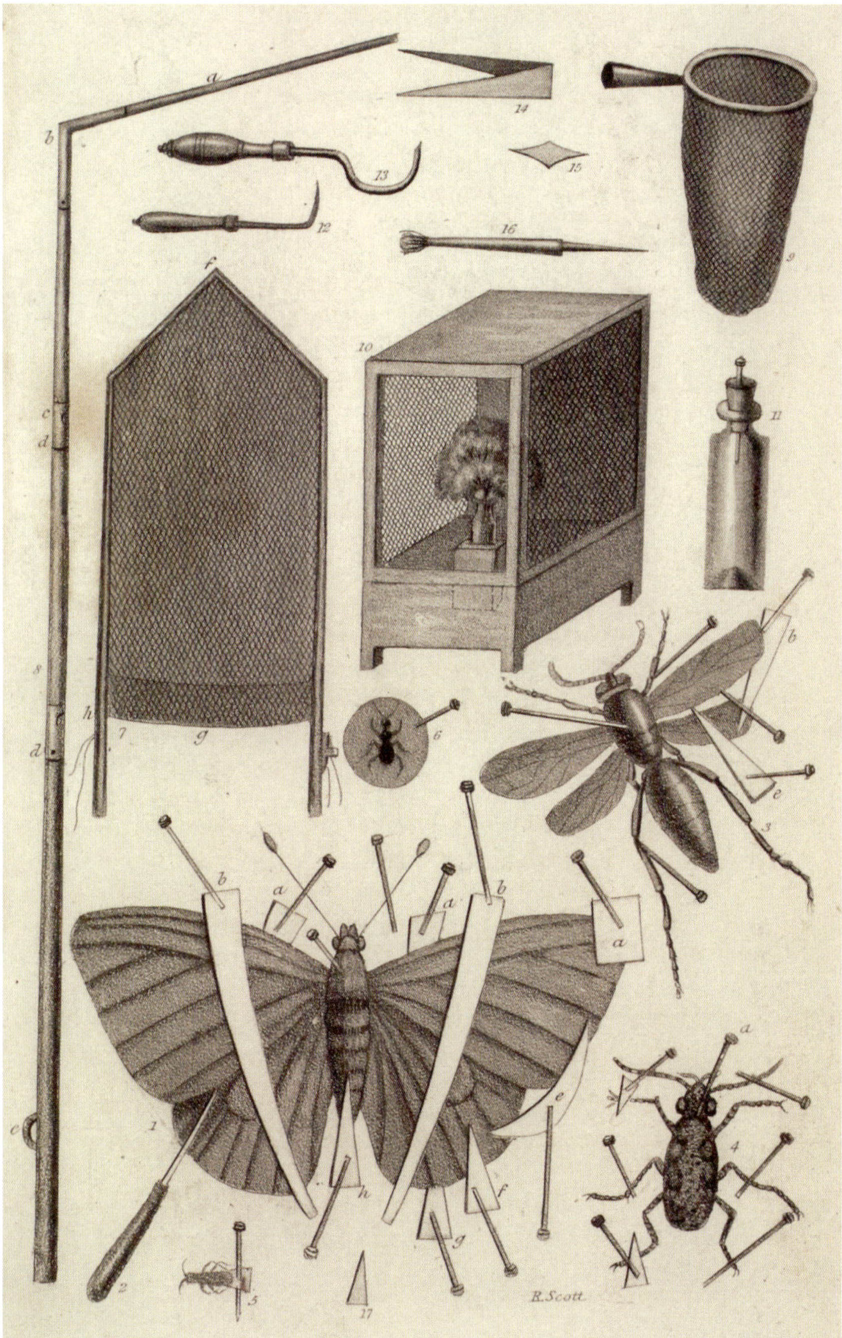

FIGURE 3.5. Thomas Brown, plate 4, *The Taxidermist's Manual, or The Art of Collecting, Preparing and Preserving Objects of Natural History* (1833), Smithsonian Archives and Libraries. Figure 7 is similar to a bat-folding net. Figure 10 is a breeding cage.

Caterpillars, on the other hand, require an entirely different approach, one informed by knowledge of their habits and probable whereabouts. Donovan provides his readers with some basic information about how to find caterpillars: "Most Caterpillars lurk among the lowest herbage in the night, begin to ascend early in the morning, and about noon are found feeding on the tops of their respective plants; they descend as the sun declines, and at the close of the evening are again concealed in the low herbage: night feeders ascend in the evening, and descend as morning approaches."[18] He notes that "numbers of the Moth tribe" feed at night, coming up from the earth, where they hide during the day. It is for this reason that "some Aurelians have sought for Caterpillars by the light of a candle or lanthorn, and have been very successful: the most valuable insects have been discovered by this means."[19] Although bat-folding nets were "intended to catch Moths on the wing," Donovan suggests that they could be used to collect caterpillars in the following manner: "Expand the net immediately under the

FIGURE 3.6.
Moses Harris, caterpillar escaping from a chip box, engraving, plate XIX, cropped, *The Aurelian: or, Natural History of English Insects; namely Moths and Butterflies* (1766), University of Alberta Library / Biodiversity Heritage Library.

bush, or branch you mean to examine, another person may beat or shake the bush with a stout stick, and not only a number [of] Caterpillars will fall down, but many of the minuter kinds of coleopterous, and other Insects also."[20] Once caught, caterpillars were put in small containers known as chip boxes and carried home, where they were transferred to larger cages.

Raising lepidopterans involved not only catching caterpillars and putting them in what were called breeding boxes but also feeding them for a few weeks as they grew larger and readied themselves for the pupal stage, when they retreated into a chrysalis or pupa to undergo transformation into a winged creature. Explaining what breeding boxes were, Lettsom tells his readers to put caterpillars "into boxes covered with thin canvas, gauze, or cat-gut, and . . . feed them with the fresh leaves of the tree or herb on which they are found." Samouelle's *Directions* emphasizes the importance of a "constant supply of fresh plants," which should be kept alive in the cage by pacing a tube in the cage to hold a bottle of water "in which the stems of the plants should be put" (figure 3.5).[21] Also, dirt mixed with wood chips or "dust from the inside of rotten trees" should also be put in the bottom of the cage for the use of "such larvae as construct their cocoons of rotten wood," as Kirby and Spence suggest.[22] Lettsom explains this process in detail:

> It may be necessary to observe that in breeding these kinds of insects, some earth should be put into the boxes, as likewise some rotten wood in the corners; because, when the caterpillars change into the pupa, or chrysalis state, some go into the earth, and continue under ground for many months before they come out into the moth state; and some cover themselves with a hard shell, made up of small pieces of rotten wood. Hence also, as many go into the earth, valuable insects may sometimes be found by digging after them a foot deep, about the roots of trees, shrubs, and plants.[23]

In *The English Lepidoptera, or, the Aurelian's Pocket Companion* (1775), a handy guide to collecting and raising English butterflies and moths, Moses Harris warns that the soil at the bottom of the breeding cage must be kept cool and moist:

> In the preservation of the Chrysalides during the winter, be sure to keep them in cold and moist places, as in a cellar or out house, otherwise the greatest part of them will be killed, especially those whose nature it is to change in the earth, as they cannot bear to be kept warm and dry, which is apt to exhale the nutritive moisture from them, hardens their shell, and weakens the insects so much, that at the time when they should appear in their flying state, they have not strength left sufficient to burst open the Chrysalis and

> come forth of their confinement; in this case the miserable animal must die, as many have done, occasioned intirely by this mismanagement of them.[24]

"Went to ground" is Abbot's phrase for the caterpillar burying itself in the dirt at the bottom of the breeding box, a phrase that appears repeatedly in his notes published in *Insects of Georgia*. He was a master of the techniques described in these manuals and routinely provided hundreds of caterpillars from different lepidopterous species with an environment that supported their growth and transition to the winged state.

Raising butterflies and moths, then and now, requires paying attention to the needs of caterpillars. They eat a lot, growing rapidly and exponentially. Hungry caterpillars, however, tend to be fussy eaters, consuming the leaves of only a few particular plant species. One of the most challenging parts of breeding lepidopterans involved figuring out what to feed the caterpillars. This was especially difficult for someone like Abbot, who in the 1770s was new to the American South and had to learn gradually through trial and error what to feed his caterpillars. Aspiring lepidopterists were encouraged to take note of what plants their captive caterpillars were found on or under, since these plants were most probably their food source. Caterpillars, Donovan cautions, "are rarely found on plants which do not afford nourishment to their species, but it sometimes unfortunately happens that stragglers are taken on some particular herbage, altogether of a different nature to its proper food; and indeed in some cases the most skillful practical entomologists are deceived, the Caterpillar refuses to eat of the proffered plant, and dies."[25]

Abbot, one of the most skillful practical entomologists of his era, did occasionally suffer the loss of caterpillars, some of which died in captivity, refusing to eat what he had given them. Not only was the aforementioned native North American silk moth, the cecropia moth, "hard to raise in the house, as it will not feed freely when confined," but the imperial moth, or as Abbot calls it, the "Great Sycamore Moth," "is the hardest one to raise of any. The Caterpillar will not feed when confined, & they commonly die in Chrysalis."[26] Abbot discovered that when caterpillars were in the pupal stage, they could be quite vulnerable, and especially if they overwintered, they were more likely to die than those caterpillars who pupated and emerged in the spring and summer months. Of the "Chestnut-coloured Underwing Moth" he wrote: "The caterpillar is not unfrequent in Autumn, but is very hard to breed, almost always dying in the Chrysalis state during the Winter."[27] He was equally distressed when the caterpillars of the "Fringe-Tree Hawk-Moth," having pupated at the end of July and buried themselves in the dirt of a breeding cage, did not make it through to the next spring: "They all died in Chrysalis during the Winter and disappointed all my hopes."[28]

FIGURE 3.7.

Abbot, "Great-Plane Tree Moth, *Phalaena Imperatoria*," imperial moth (*Eacles imperialis*), original drawing, TAB. LV, Smith and Abbot, *Insects of Georgia* (1797). Courtesy of John Work Garrett Library, Sheridan Libraries, Johns Hopkins University.

FIGURE 3.8.
Abbot, "Snail Tussock," Abbot's bagworm moth (*Oiketicus abbotii*), close-up of caterpillar's "house," watercolor drawing, plate 96, "Moths of Georgia" (ca. 1817–20), Hargrett Rare Book and Manuscript Library, University of Georgia Libraries.

FIGURE 3.9.
Abbot, "Bat-Caterpillar Moth, *Phalaena Pithecium*," hag moth (*Phobetron pithecium*), original watercolor drawing, TAB. LXXIV, Smith and Abbot, *Insects of Georgia* (1797). Courtesy of John Work Garrett Library, Sheridan Libraries, Johns Hopkins University.

Inclement weather also impacted his ability to collect and raise Lepidoptera and other insects. Explaining to a client why he could not send him "a larger variety" of specimens, Abbot lamented in November 1821:

> This has been a bad year for collecting [with] late and severe frosts in the Spring, that killed all the fruit and blossomes, to which suceeded an uncommon wet Summer that always kept the Swamps and low ground under water. And I must own I did not pay that Attention to Collecting in the spring, which I shou'd had I received your Letters first. The Sphinxes are in general rare. I never but once met with 3 Kalmiae Catterpillars from which I bred 2 flies. . . . I did not br[e]ed one Cecropia this Year, but have 4 now in Chrysalis, and a number of others.[29]

He complained to his correspondent, the entomologist Thaddeus W. Harris, that the winter of 1834–35 had been so cold that the green sphinx chrysalid had

FIGURE 3.10. Abbot, "Muscadine Sphinx, Sphinx Crantor," achemon sphinx (*Eumorpha achemon*), original watercolor drawing, TAB. XLI, Smith and Abbot, *Insects of Georgia* (1797). Courtesy of John Work Garrett Library, Sheridan Libraries, Johns Hopkins University.

died in his house.[30] Abbot's lament about the loss of this green sphinx suggests that despite being in his eighties, he was still raising caterpillars and continued to be deeply immersed in the study of insects.[31]

Abbot knew a great deal about the caterpillars he raised. He knew what to feed them and where to find them. "I have found them spun up in the River Swamp on Large Sycamores under the loose Bark of the Tree," referring to a tussock moth.[32] He also studied their behavior and took an interest in how they interacted with their environment. The "Snail Tussock" (*Oiketicus abbotii*), which apparently ate a range of plants, made ingenious use of leaves and stalks to make "itself a house" (figure 3.8):

> This singular Species feeds on the Red haws, Sassafras, Persimmon &c. It spins itself a house in which it lives hanging it to the stalk and feeds by drawing the leaves to it like figure 1. It fastens pieces of the dry stalks on the outside and enlarges it as it grows bigger, fig. 2 shews the Caterpillar taken out of its house. It closed up the opening and changed into Chrysalis 16th June, bred 5th July. fig. 3 is the Chrysalis of the female, and fig. 4 of the Male. [F]igs 6 and 7 shews 2 views of the female, resembling a Maggot, having not the least rudiments of wings. The Caterpillar is frequent, but the Male very rare, and flies very swift.[33]

A little brown moth's larva behaved in ways that caught his attention:

> This Catterpiller is green with a dark green longitudinal line on the back; head yellowish green. It was taken folded up in the leaves of Lilac the 4th of June. The 8th it changed into a Chrysalis of a dark brown, darkest on the thorax and wings. Out of which the 22d. of June came the Moth. See the 26 Drawing figure 148. These Catterpillers are very nimble in their motion both forwards and backwards, as I have observed most of them that fold themselves up in the leaves. If you force them out of their Leaf[,] they let themselves fall by a thread to the next leaf or branch, by which when all is quiet, they climb up to their former habitation.[34]

Describing similar behavior of what he calls the "Chestnut-bordered Under-wing," he observes that the caterpillar folds muscadine or wild grape leaves around itself, forming a "retreat, or house." Then in May it spun up "on the surface of the ground in a case of dirt," where it remained in this pupate state until the following March. These moths were "most frequent near the great Ogeechee River."[35]

He also paid close attention to the often unusual appearance of his cater-pillars, remarking, for instance, on the "Singular shape" of the larva of the "Bat-

Caterpillar Moth," known today as the hag moth (*Phobetron pithecium*). Its caterpillar, the "monkey slug" in modern parlance, has "a kind of furse [gorse—a thorny plant] growing on its back resembling wings or horns. When it spins up these fall off & it mixes them with its Web."[36] The care with which Abbot recorded the changes that caterpillars underwent as they grew and molted, shedding their skin as it became too tight to contain their swelling bodies, is evident in his description and drawing of the caterpillar of the "Muscadine Sphinx." "When small," meaning before its final molt, "the Catterpillar has a horn on its tail, which turns back over its back like a Dogs tail. When [they] shed their skin the last time, they shed also their tail, only having afterwards a small protuberance, & changes their colour."[37] The illustration of the "Muscadine Sphinx" depicts this process of change and growth with two caterpillars in different stages of molting, a smaller one with the dog's tail, and a larger one with a small bump where the horn had been (figure 3.10). Abbot noticed that a caterpillar's appearance is affected not only by the molting process, changing according to which instar it is in, but also by the food it eats. The caterpillars of the "Orange Brown-Barred Moth," for instance, vary in color depending on whether they eat walnut leaves, which leave them "black with white hairs," but "when on Oak they are more yellow, like the English Buff Tip Catterpillar."[38] These notes testify to the very high level of scrutiny and engagement that Abbot brought to the study of Lepidoptera and the project of raising caterpillars.

The narrative most frequently told in his notes is the story of metamorphosis, beginning with a caterpillar that has been caught often on the plant it eats; then, depending on the kind of lepidopteran it is, it either spins a cocoon or "changes" into a chrysalis; finally, after a couple of weeks or several months, it emerges as an adult "in its winged state."[39] The note for the "Black and Yellow Swallow-Tailed Butterfly" is typical for a butterfly entry: "The Catterpillar of this species eats fennel & Rue. Changed into a Chry.[s] 12th July. The Butterfly was bred the 20th."[40] Typical for a moth entry is the following describing the "Oak Beauty": "Was taken feeding on the Black Oak, it eats also other kinds of Oaks. It spun upon the Surface of the Ground 22[d] of May. Bred 13th of February."[41] Most narratives begin with the caterpillar that has been "taken" on its host plant, but occasionally Abbot mentions hunting for eggs laid by butterflies and moths. With the "Swallow-tailed Skipper Butterfly," Abbot was more successful finding its eggs than hunting for its caterpillars, since "the Catterpillars of all the species of *Skippers* fold the leaves together for safety like the English Admirable, which makes them not easy to be discovered."[42] Some skippers and moths were often easier to find when in the chrysalis state than as larvae. The chrysalis of the "Great Black and White Tussock Moth," for instance, "may be often met spun up to weeds & Bushes fastened to a twig or stalk as in the figure."[43] Once a caterpillar has been found and collected, it is put into a breeding box, where the caterpillar is fed until it retreats into a chrysalis or a cocoon to undergo preparation for the final stage of its transformation (figure 3.11).

Abbot describes this moment of transition into the pupal stage with plain-spoken descriptions such as "tied itself up," "suspend itself by the tail," "went to ground," and "bury itself." For a more detailed and dramatic portrayal of this stage, we can turn to Moses Harris's description of this process. When caterpillars are "full fed," they make themselves a "secure retreat," where over the course of a few days

> they shrink and grow shorter, losing the use of their feet entirely, and appear as if in great agony; at length the skin on the two first joints, behind the head, which at this time appear very much swelled, bursts, or rather slits, and opens some way down the back and cross the head, that in some you would, at first sight, suppose the head of itself was divided. During this time, the caterpillar strives to throw off its skin, which, however, he facilitates by a motion very peculiar, working off the skin joint by joint, till it arrives quite to the tail; nor does it cease twisting and turning itself till quite disengaged from it: it is very soft and tender; and it is generally a day before the shell of the chrysalis becomes hard. . . . [W]hen the shell becomes hard, it lies motionless, unless disturbed by some accident, till the expiration of a certain time, and at length breaks forth into the winged state: but during the time of its being in the chrysalis state, it receiveth no nourishment of any kind, although some remain in that state near two years.[44]

Although Abbot's notes do not rise to the stylistic level of Harris's eloquence, Abbot shared Harris's deep fascination with metamorphosis, as evidenced by his having depicted and described caterpillars from over three hundred different North American lepidopteran species.

The details that appear in Abbot's notes about the various stages of metamorphosis were most likely recorded initially in a journal. It was commonplace for those who raised butterflies and moths to record and store data on these insects' stages of transformation in what was called a breeding journal. While none of Abbot's journals have been located, one of John Francillon's breeding journals has survived and is currently housed in the library of the Royal Entomological Society in St. Albans, England. Breeding journals are one of those unsung Enlightenment paper technologies that helped butterfly and moth collectors and breeders to gather, organize, and store information about raising caterpillars.[45] Francillon recorded when and where he caught his caterpillars and on what plant he found them. His journal is a bound volume with originally blank pages on which he drew grids, looking not unlike today's digital spreadsheets.

FIGURE 3.11.
Abbot, "Great Black and White Tussock Moth, *Phalaena Velleda*," large tolype moth (*Tolype velleda*), with a chrysalis attached to a twig, original watercolor drawing, TAB. LII, Smith and Abbot, *Insects of Georgia* (1797). Courtesy of John Work Garrett Library, Sheridan Libraries, Johns Hopkins University.

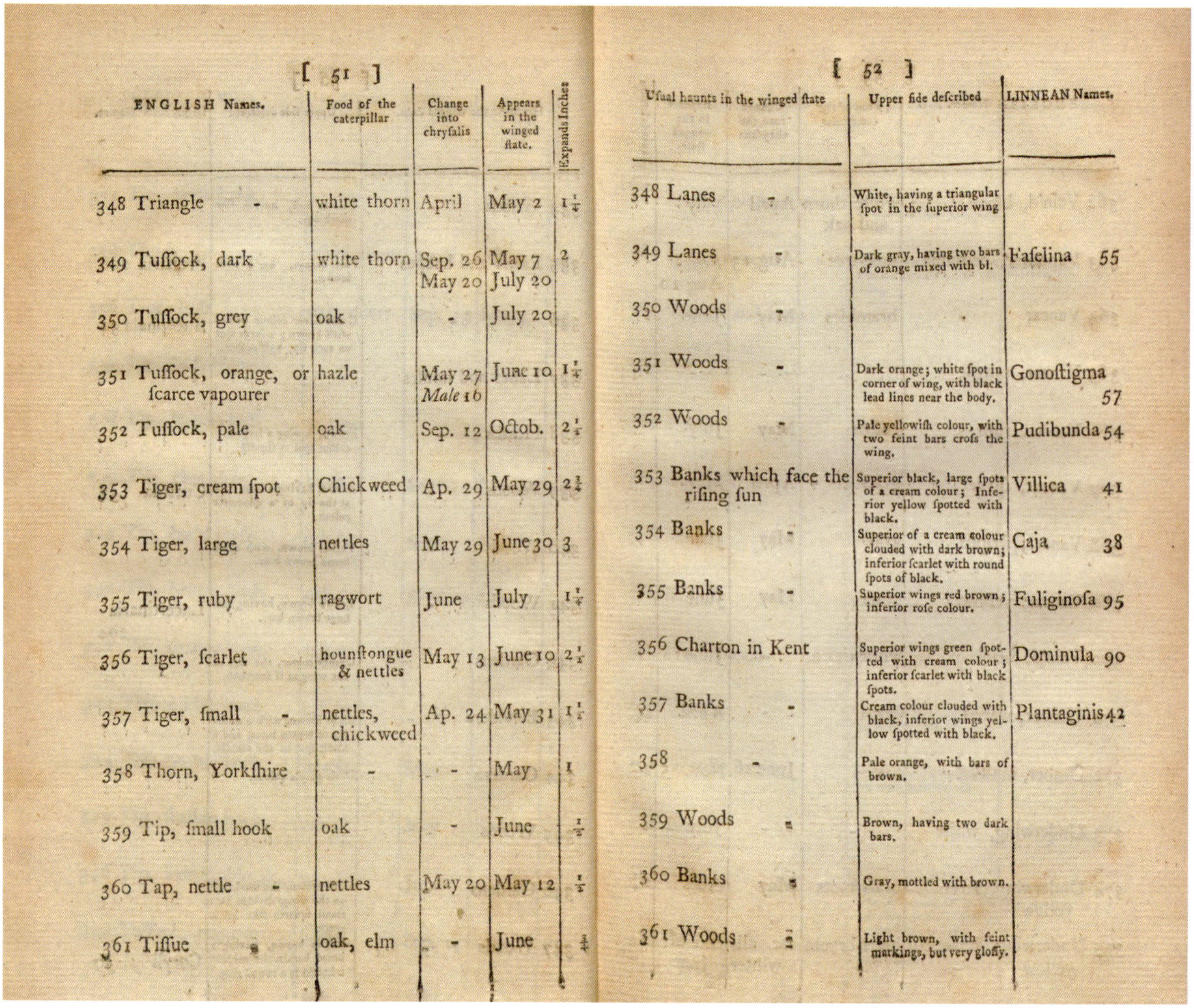

[51] ENGLISH Names.	Food of the caterpillar	Change into chryſalis	Appears in the winged ſtate.	Expands Inches	[52] Uſual haunts in the winged ſtate	Upper ſide deſcribed	LINNEAN Names.
348 Triangle -	white thorn	April	May 2	1¼	348 Lanes -	White, having a triangular ſpot in the ſuperior wing	
349 Tuſſock, dark	white thorn	Sep. 26 May 20	May 7 July 20	2	349 Lanes -	Dark gray, having two bars of orange mixed with bl.	Faſelina 55
350 Tuſſock, grey	oak	-	July 20		350 Woods -		
351 Tuſſock, orange, or ſcarce vapourer	hazle	May 27 *Male* 16	June 10	1¼	351 Woods -	Dark orange; white ſpot in corner of wing, with black lead lines near the body.	Gonoſtigma 57
352 Tuſſock, pale	oak	Sep. 12	Octob.	2½	352 Woods -	Pale yellowiſh colour, with two feint bars croſs the wing.	Pudibunda 54
353 Tiger, cream ſpot	Chickweed	Ap. 29	May 29	2¾	353 Banks which face the riſing ſun	Superior black, large ſpots of a cream colour; Inferior yellow ſpotted with black.	Villica 41
354 Tiger, large	nettles	May 29	June 30	3	354 Banks -	Superior of a cream colour clouded with dark brown; inferior ſcarlet with round ſpots of black.	Caja 38
355 Tiger, ruby	ragwort	June	July	1¼	355 Banks -	Superior wings red brown; inferior roſe colour.	Fuliginoſa 95
356 Tiger, ſcarlet	hounſtongue & nettles	May 13	June 10	2½	356 Charton in Kent	Superior wings green ſpotted with cream colour; inferior ſcarlet with black ſpots.	Dominula 90
357 Tiger, ſmall -	nettles, chickweed	Ap. 24	May 31	1½	357 Banks -	Cream colour clouded with black, inferior wings yellow ſpotted with black.	Plantaginis 42
358 Thorn, Yorkſhire	-	-	May	1	358 -	Pale orange, with bars of brown.	
359 Tip, ſmall hook	oak	-	June	½	359 Woods -	Brown, having two dark bars.	
360 Tap, nettle -	nettles	May 20	May 12	½	360 Banks -	Gray, mottled with brown.	
361 Tiſſue -	oak, elm	-	June	¾	361 Woods -	Light brown, with feint markings, but very gloſſy.	

FIGURE 3.12. Moses Harris, *The English Lepidoptera, or, the Aurelian's Pocket Companion* (1775), Library of the Netherlands Entomological Society / Biodiversity Heritage Library.

The format of Francillon's journal owes much to Moses Harris's *Aurelian's Pocket Companion*, a small book that popularized collecting caterpillars by providing information about when, where, and how to find caterpillars of nearly four hundred different lepidopteran species. Each page of the book had a printed grid made up of categories, columns, and the common names of English Lepidoptera, and the empty spaces awaited the penciling in of information about where caterpillars were caught, what they probably ate, and what the moth or butterfly that emerged from a chrysalis or pupa looked like. A column for the English common names of butterflies and moths was on the far left, followed by seven columns that provided information and space for noting (1) the kind of plants "Insects feed upon in the Caterpillar state"; (2) "the times of the transformation into the Chrysalis"; (3) when "it appears in the winged state" and how long it was in the chrysalis state; (4) measurement of the wings of a "Fly" from one to another; (5) "places where the insect frequent, or are commonly found, in the winged state"; (6) "a short description of the upper side of the wings, intended to assist the memory of the reader"; and (7) the "Linnean Names" and the number Linnaeus assigned to each insect in his twelfth edition.[46] The copies of this book that I have seen have been annotated by their owners, making this "compendium," as Harris termed this small volume that contained "every new discovery of the author's researches," into a repository for its readers' discoveries as well.[47]

The format of Harris's *Aurelian's Pocket Companion* was modeled on Harris's own breeding journals, one of which has survived and is held at London's Natural History Museum. This multimedia affair was dedicated to studying and recording the caterpillars he found and raised. Harris pasted small, cutout drawings of caterpillars and moths onto the pages of this hand-ruled volume, making for a lovely though somewhat chaotic collage of notes and drawings. Pasted onto the recto page, for instance, is a charming drawing of a yellowish green "Bee Tyger" caterpillar (*Acherontia atropos*, a sphinx moth) "found at Stepney by

FIGURE 3.13. Moses Harris, engraving, plate XXXVII, *The Aurelian: or, Natural History of English Insects; namely Moths and Butterflies* (1766), University of Alberta Library / Biodiversity Heritage Library.

farmer Patten feeding on a field of potatoes July 29, 1778." Near a small drawing of a chrysalis Harris wrote: "This is a true Drawing of the Chrysalis of a Bee Tyger drawn ye [the] 17th of August 1778 [undecipherable] Mos Harris." Across from these drawings are penciled notes squeezed into one of the boxes made by the grid with the number 152 in the farthest left vertical column: "This large Catr. was sent me by Fr [Farmer] Patten July 29 nearly tufted. The whole was a deep Yellow Green. The Stripes on his sides purple. Ye [The] breathing holes black. Went down in 3 Days 1778. Came forth Octo[be]r 8 a Cripple."[48] Apparently, this caterpillar's metamorphosis was not completely successful; perhaps it had been damaged during its capture or during its transport from Stepney to Harris's residence. This breeding journal contains drawings and information that Harris then used for his magnificent book *The Aurelian: or, Natural History of English Insects; namely Moths and Butterflies together with the Plants on which they Feed* (1766, 1778), and this particular caterpillar and its hawkmoth imago, also known as the death's-head hawkmoth, are depicted in plate XXXVII, dedicated to Dru Drury.

Abbot followed Harris's practice of keeping breeding journals for information to be mined for later use. The notes that Abbot wrote for *Insects of Georgia* were based on journal entries, as were his notes that accompanied the drawings he sent to Francillon, Swainson, LeConte, and others. Although Abbot may have used a template to reproduce drawings of a particular species of a butterfly (more on this in the next chapter), this does not necessarily mean that the accompanied notes were also copied from a template. The drawing would have represented an ideal type; therefore, it would have made sense for Abbot to develop an image that satisfied the taxonomic requirements and to reuse that image as a template for further drawings. However, the notes seem to be about individual insects; there are, in some cases, significant discrepancies between the notes that accompanied drawings of the same lepidopteran species. For example, the notes on "A Great Brown Emperor Moth" in Francillon's volume 17 state: "This caterpillar feeds on the Wild Plum, Apple Redroot etc. It spun up 3rd September in a web fastened to the branch of a Tree. The chrysalis is enclosed in an Inner Web resembling that of a silk worm. The moth was bred 16th March."[49] Comparing these dates with those in the notes describing the "Great Brown Emperor Moth" depicted in Smith and Abbot's *Insects of Georgia* (1797), which were probably written a decade before those in Francillon's collection, we get June 17 as the date it spun up, March 30 as the day it emerged from the chrysalis, and a rather poignant statement about how difficult it is to raise this caterpillar, as "it will not feed freely in confinement" (figure 2.12).[50] These two sets of notes describe different moths of the same species. In this case as well as others, Abbot seems to be referring to individual moths, which is extraordinary, especially in this era when scientific focus was on representative types, not individuals and their differences.

Using Abbot's notes that accompanied his drawings of lepidopterans, we can speculate about how many species of Lepidoptera Abbot might have raised. Abbot raised all the butterflies and moths depicted in the two volumes of *Insects of Georgia*, amounting to 104 different species. Likewise, the notes for volumes 16 and 17 of Francillon's manuscript collection indicate that Abbot raised all the butterflies, moths, and beetles that appear in the 277 plates of these two volumes.[51] Raising so many different species of Lepidoptera (plus a few species of Coleoptera) even over the course of a decade or two would imply a fairly large operation with many breeding cages for each species. According to Donovan, breeding cages should contain only one species, since caterpillars were liable to eat caterpillars of another species and "if left without food, will devour those of their own kind also."[52] If Abbot had a cage for each species of Lepidoptera, the number of breeding cages he must have had suggests an undertaking that occupied a significant amount of space, possibly a whole structure or two to shelter the caterpillars from Georgia's hot summer sun. Keeping caterpillars fed with the right plants must have been labor-intensive, especially if he were raising more than a few of one species at a time, all of which suggests the possibility that the enslaved men and women who worked for Abbot may have built the cages and fed the caterpillars with plants they had gathered.[53] One way or another, Abbot raised thousands of lepidopterous and coleopterous specimens, but this was only the first step in turning living creatures into scientific specimens. The caterpillars that had been raised with such care and had metamorphosed into butterflies and moths eventually were killed, preserved, and shipped up north or abroad to Europe to fill the cabinets of entomologists and natural history collectors who were eager to obtain these exotic insects from the American Southeast.

Killing Insects

The particular skills that made Abbot such an expert specimen collector were those involved in killing and preserving insects as specimens. These skills can be inferred from how others described these practices in sources such as Drury's *Directions*, which he gave to his agents; Harris's introductory remarks to his stately *Aurelian*; and Donovan's booklet *Instructions for Collecting and Preserving Various Subjects of Natural History*."[54] Drury advised those who collected on his behalf that the best way to kill a moth or a butterfly was to merely squeeze its abdomen or use a pin to pierce its thorax.[55] Donovan, a natural history collector and illustrator and proud owner of Abbot's specimens, cautioned collectors that killing moths and butterflies caught in a net was not always as simple as Drury suggested in his *Directions*. For some larger moths this method did not work very well, as Donovan explains:

> For small Moths it is only necessary to put the pin through the thorax and they die in a very short time; but for larger kinds, the pin should be dipped in strong aqua fortis [nitric acid] before it is put through the Insect. It is very difficult to kill the largest kinds of Moths and Sphinxes:— select a large pin (comparatively for the size of the Insect) and dip it into aqua-fortis as before, but immediately that the pin is forced through the thorax withdraw it, and put a drop of aqua fortis into the wound; should this prove insufficient to kill it, put the point of the pin through a card, and hold it in the flame of a candle until it becomes red hot; this will kill the Insect immediately, and the card will protect it from being injured by the flame.[56]

Dragonflies were also notoriously difficult to kill, as "they are extremely tenacious of life"; Donovan attests to having seen a dragonfly live two days "on the pin." He suggests: "The most expeditious method of killing those creatures, is to run a red hot wire up the body and thorax, for they will live a considerable time in agony if you attempt to kill them with aqua-fortis as before directed for the Moth tribe."[57] Once dead, insects needed to be "set," which means that the wings of butterflies, moths, wasps, dragonflies, and other winged insects had to be spread open and held with braces and pins in that position until dry. Harris explains this technique for the novice Aurelian:

> Having returned home with your insects, look in your boxes, and observe which of them is fit to set, such as are dead, but not stiff, are so; then proceed to manage with them as follow[s]: Take a fly, and observing if the pin be perpendicularly run through the body, place it on the setting board, then with your point gently raise one of the upper wings, till such time as the tip be even with the nose of the fly; this done, fix one of your card braces on that wing, to prevent its giving way; do the same by the wings on the other side, and your fly will be properly extended; let the braces remain on the wings of butterflies a fortnight, on those of large moths a month.[58]

Donovan cautions that setting insects must be done fairly soon after their death: "All Insects must be set while they remain limber, for if the parts stiffen they are apt to snap; they may be relaxed by floating them in a pan of water." Preserving large moths requires a bit more finesse, since they have "entrails" that do not simply dry out but rot instead. He advises: "Large Moths are to be opened in a straight line along the under side of the body, the entrails, &c. taken out, and the cavity filled with fine tow or cotton.—*Note*, This should be performed soon after the Insect is dead. The most delicate specimens may be preserved entire by this means; we have some very valuable Sphinges, Moths, &c. which were collected by an intelligent person in *North America*; they retain their colours to the utmost degree of perfection, and have only been prevented from changing black by this simple preparation."[59] This "intelligent person" must certainly be Abbot. Clearly, Donovan thought quite highly of Abbot's skills as a practical entomologist.

Donovan's instructions for capturing and killing insects also include a few passages about the ethics of certain methods used to kill insects. Not until the 1790s did questions about cruelty to animals arise within the Evangelical movements and within the new cult of sensibility.[60] Donovan published his instructions in 1794 and may have been influenced by these emerging anticruelty discourses. He was quite critical of one particular method of killing caterpillars that involved roasting a live caterpillar over a fire while squeezing out the caterpillar's internal organs through its anus: "This is a most cruel operation on the little victim, and such as must shock the feelings of the human soul; if therefore any other method can be introduced which will effect the purpose in a short time, the practice should be exploded as wanton barbarity."[61] Another method for killing caterpillars involved immersing them in wine, a practice described by Kirby and Spence, who admired Abbot's "excellent method of preserving caterpillars, so that his specimens retain their colours and other attributes, and look as if they are alive."[62] He admits he was "not acquainted with his process, but the following will answer very well":

> The animal must first be killed by immersion in spirits of wine; next you must eviscerate it, which is best effected by gradual pressure of the finger and thumb. You must begin at the head, and so proceed till all the fluid contents of the body have passed out at the anus, which you may enlarge with a fine pair of scissors, being careful not to injure the anal prolegs. When you have cleared the skin as much as possible, introduce a fine glass tube, or a piece of hay or slender straw into the anus, round which, as near to the extremity as may be, pass loosely a fine thread; then blowing through the tube, when the skin is fully inflated, withdraw it, at the same time pulling the thread tight and securing it by a knot. The caterpillar will now exhibit its proper shape and colours; to retain which, all that is necessary is to hold it near the flame of a lamp until perfectly dry, which will be in a few minutes, when it may be placed in the cabinet along with the image to which it belongs.[63]

Donovan questions whether drowning caterpillars in wine and other spirits was effective, since some seemed to be able to live up to "eight days with only two of its anterior spiracula in the air. (Spiracula or spiracles are pores on the caterpillar's body that enable it to breathe.) He recommends instead to take the caterpillar when it is at "the point of casting its last skin," drop it into scalding water, and immediately withdraw it. "The creature will be killed instantly" and after a vinegar bath will be ready to eviscerate, removing the skin in the process, which then, like the above method, is held over a fire to dry out as it is being "blown up by means of a glass tube," a technique along the lines that Kirby and Spence suggest Abbot might have used.[64]

Donovan's concerns about what he thought were cruel methods of killing caterpillars was taken up nearly a decade later by Adrian H. Haworth in his *Lepidoptera Britannica* (1803). He expressed disapproval of how some lepidopterans, large moths in particular, were being killed:

> The larger Moths are very tenacious of life, and difficult to kill without spoiling them: the usual way of compressing the *thorax* is not sufficient: they will live several days after the most severe pressure has been given there; to the great uneasiness of any humane Aurelian [butterfly collector]. The methods of suffocation by tobacco or sulphur are equally inefficacious, unless continued for a greater number of hours than is proper for the intended preservation of the specimens. Another mode now in practice is better; and however fraught with cruelty it may appear to the inexperienced collector, is the greatest piece of *comparative mercy* that can, in this case, be administered. When the larger Moths must be killed; destroy them at once, by the *insertion of a strong red-hot needle into their thickest parts beginning at the front of the thorax*. If this is properly done, instead of *lingering through several days, they are dead in a moment*.[65]

Haworth, who shared some of Donovan's concerns about collecting practices being potentially cruel, argued, however, that invertebrates have nervous systems that are very different from those of vertebrates and, as a result, could be understood as not experiencing pain the same way vertebrates do. Although Haworth quotes Shakespeare's lines about the poor, tread-upon beetle who "In corporal sufferance feels a pang as great / As when a giant dies," he begs to "differ from our great poet on this particular point, because the difference may ease the humane Entomologist of many a rising qualm." Arguing that insects "are not near so susceptible of the sensation we call pain" and that they "have not such acute organs of feeling pain," he concludes:

> If insects were in reality possessed of the degrees of feeling pain, that have been erroneously attributed to them, they would never survive the loss of limbs forcibly torn off, as often happens to fighting Crabs, fighting Spiders, and Tipulae [flies] self-extricated from children's hands, or spider's webs, with the dear-bought ransom of their joints. Neither would common Flies live long without their heads, wings, or other parts; or Coleopterous Insects just escaped from the perforation of an Aurelian's pin, attempt to copulate with their mates: —yet all these things I have myself been an eye-witness to their doing. These facts, and these remarks, are surely capable of removing the

> numerous charges of cruelty, which have hitherto attached to the Collectors of Insects.[66]

Today insects collected by entomologists are usually instantly killed in the field when exposed to a gas (ether, chloroform, or ethyl acetate) in a kill jar. Some lepidopterists freeze the butterflies and moths they catch, which also kills them in a purportedly painless way. Questions about how much pain insects experience have yet to be resolved, and entomologists are in the process of doing research on what constitutes insect pain and suffering.[67]

Preserving Specimens

Abbot sent dried and pinned insects and taxidermized bird skins to Europe in specially built boxes designed to protect specimens from damage during travel.[68] Even carefully packed specimens traveling relatively short distances over land would be jostled and jolted as they made their way over rough dirt roads, often resulting in torn butterfly wings, lost beetle antennae, and missing spider legs. Travel from one side of the globe to the other involving rough seas multiplied the chances of even greater damage, compounded by saltwater, damp air, and aggressive customs officials. Perhaps most damaging of all were insect infestations that could wipe out whole collections of bird and insect specimens. While fish and reptile specimens, stored in jars filled with alcohol, wine, and other liquid preservatives, were safe from these destructive vermin, bird skins would be consumed unless they had been dosed with arsenic, which had serious side effects for those who prepared and handled these specimens. Insects pinned to cork boards fell victim to insect infestation, leaving specimen hunters like Abbot searching for remedies to battle these destructive forces.

The key to success as a supplier of natural history specimens was (and still is) the ability to keep dead animals, specifically birds and insects, from being eaten by vermin. In the notes accompanying a drawing of a mantis, Abbot tells how his mantis specimens were gobbled up by what he calls maggots:

> Female. Taken 6th Sept. in Oak Woods and low Grounds of Savannah River. The male flies in the even'g; its body is slenderer and wings longer than the Female here figured. . . . I have taken several Females seemingly in full health and strength, and pinned them in my Box[.] Upon opening the Box a few hours after, have found several large maggots full fed that had eaten themselves out of the abdomen and eat all the inside, having been bred in the Mantis before it was taken. . . . [N]ot uncommon, called in Savannah Walking leaf.[69]

The maggots Abbot describes here are the larvae of an insect that uses the body of the mantis to house and feed its offspring while in the larval and parasitic stage, a fairly common occurrence (parasites can even have parasites) but one that speaks to the complexity of the insect world. The maggots Abbot found may have been the offspring of an ichneumon wasp, a parasitoid that lays its eggs on or inside another insect's body, which is then eaten by the wasp's hatching larvae. Typically, the wasp lays its eggs in caterpillars or beetle larvae, but in this case, the wasp chose a praying mantis. Although Abbot was surprised and disappointed that these parasitoids had managed to destroy his mantis specimens, that did not stop him from collecting and drawing the wasps, noting where and when he caught them.

These parasitoids were so morally repugnant to Charles Darwin that their existence caused him to question his religious convictions. Writing to fellow naturalist Asa Gray in 1860, he opined: "I own that I cannot see, as plainly as others do, & as I sh[oul]d wish to do, evidence of design & beneficence on all sides of us. There seems to me too much misery in the world. I cannot persuade myself that a beneficent and omnipotent God would have designedly created the Ichneumonidae with the express intention of their feeding within the living bodies of Caterpillars, or that a cat should play with mice."[70] Far less squeamish than Darwin, Moses Harris was fascinated by Ichneumonidae and drew hatching ichneumon wasp larvae emerging from a caterpillar, describing this moment thusly:

> An Ichneumon Fly often lays its Eggs in the Body of the Burnet Caterpillar, which are there hatch'd, and produce small white Maggots, which feed on its Flesh. The Caterpillar, not withstanding it is infested with so many of these Devourers, that they swell his Body much beyond its natural Bulk, sometimes lives till the Time they generally begin to spin their Web, in which they inclose themselves when about to go into Chrysalis, but is destroy'd by these Insectophagi before he can complete it. When these Maggots are near full fed, they eat their Way out through the Skin of the Caterpillar, as at (m), and leave it shrunk and dying, as represented at (k). In a short Time after they go into Chrysalis (n), and in about ten Days more produce Flies like their Parent (l).[71]

Parasitoids were far less destructive to the collecting process than other insects, such as the *Dermestes* beetles, which could devour whole entomological and ornithological collections. Of one kind of *Dermestes*, Abbot wrote: "Taken and was bred in March. they continue to breed and make great havoc all the Summer among Collections of Insects, unless prevented by great care with close Boxes & Camphor."[72]

FIGURE 3.14.
Moses Harris, ichneumon wasp larvae leaving the body of a caterpillar (lower left), engraving, plate I, *The Aurelian: or, Natural History of English Insects; namely Moths and Butterflies* (1766), University of Alberta Library / Biodiversity Heritage Library.

Abbot worried, as did every specimen hunter, over what chemicals to use to preserve animal bodies. Insect specimens required special treatment, from dipping pins in turpentine to lining drawers with expensive paper and blowing tobacco smoke into the boxes in which they were stored and transported. Swainson referred to these "processes by which the form and substance of animal bodies may be preserved from decay" as "an art" that was "absolutely essential to be known to every naturalist; since, without it, he cannot pursue his studies or preserve his own materials."[73] Abbot complained to Swainson how hard it was to protect specimens from infestation: "The destroying Insects in this Country is wonderfully bad both to Birds and Insects. I have failed in every method hitherto to preserve them, every poison that I have tried being insufficient, except the Bird or Insect, could be entirely soaked in any poison, which wou'd then destroy its colours. I have not known any preserved and sent from Europe, that can be kept here.—The best preservatives I can find is Camphor and fresh Spirit of Turpentine."[74] Abbot acquired camphor and other supplies from Augustus Oemler, a Savannah apothecary and amateur botanist with whom Abbot became quite friendly.[75]

Camphor, Abbot's weapon of choice in the battle against these "insectophagi," may have been recommended to him by Drury, who often counseled those who collected insects for him in foreign locales to use camphor. Drury wrote to the Honourable Lady Anne Monson, an enthusiastic amateur naturalist who was about to embark on a trip to India, explaining how to deal with insect-destroying vermin: "When yr. Ladysh[i]p arrives in India you will find a small species of Ant ye greatest enemy to be encountered, nor will anything but Camphire secure your insects from them. It will therefore be absolutely necessary always to keep in Boxes in little muslin bags & never let them be without it[.] [A]s it evaporates the box must be supplied with fresh. The neglect of this caution may lose a whole Collection in a few days."[76] Camphor was also endorsed by Donovan, who urged collectors to use "a quantity of camphire for each drawer, or the mites will destroy the Insects," even though some collectors preferred to wash the cork-lined drawers of their cabinets with "wine and corrosive sublimate" and "moisten the paper after it is pasted on the cork with alum-water," a method Donovan thought far less effective than camphor.[77] Instead of using camphor, Harris had a different method to prevent "small Animalcule" from destroying insect collections: "The smoke of tobacco is the only effectual remedy, which must be blown through the small end of a pipe admitted through a hole made for that purpose at the back part of the drawer or box; this not only corrects the putrid and stagnant air, but destroys those formidable enemies which often destroy whole cabinets of Insects."[78] It's too bad for Abbot that Harris's remedy of tobacco smoke did not work as well as camphor, because camphor, a product derived from a South Asian tree, had to be imported and purchased from apothecary shops, whereas tobacco was widely available to him in Georgia. Turpentine, too, was readily available in Georgia, and Abbot employed it to sterilize the pins he used to pin butterflies and moths to the cork-lined drawers he packed into boxes for shipping overseas.

The pins that Abbot used had to be chosen carefully, since not just any pin would do, as his mentor Drury would have insisted.[79] Drury was adamant that his insect hunters were to use "long pins," as he believed that short pins were responsible for damaging butterfly wings.[80] "I charge you never use any of those cursed small pins again," he wrote to Smeathman while he was collecting on Drury's behalf in Sierra Leone. "Many of the Insects were destroyed by the pins breaking, the saltwater having made them rotten [corroded]."[81] Drury's outburst over "those cursed small pins" occurred in response to an embarrassing situation. Drury had persuaded Henry Seymer, a landed gentleman living in Dorset, to purchase some of Smeathman's butterflies, as it was the "only opportunity that has happened in my memory for obtaining any Insects from Africa, and if any accident should happen to deprive us of Mr Smeathman I am confident there never would be another person that would undertake the laborious Task."[82] Seymer agreed to Drury's proposition, but when these African butterflies arrived in Dorset, they were damaged. Seymer wrote to Drury complaining that the box he had received from him contained tattered butterflies. Drury was eager to protect Smeathman's reputation as an expert field collector, and he assured Seymer that the specimens he had sent from London to Dorset were "very fine" and must have been damaged on the Blandford coach's journey from the Black Bear tavern in Piccadilly to Dorset:

> I was afraid some mischief would befall them, & you have too truly verified my fear. Almost all the specimens when I put them up were very fine & therefore I have the more reason to lament the havoc that has been made. Some of the number you have mentioned were the finest I ever beheld, nor were they ye [refuse] of what had been chose by the Subscribers. . . . [M]any collectors here in London were exceeding desirous of having part of the very parcel that was sent to you & at the same price. So that on the whole I must impute your dissatisfaction intirely to the Accident that has befell them.[83]

Whether the tattered wings were Smeathman's fault or the Blandford coach's, we cannot know, but we will have to trust Drury that the wings were intact when they left London. However, the year before Drury did chastise Smeathman for sending specimens with "wings ragged—send no more of that sort it will do you harm," meaning Smeathman's reputation would have been damaged in London if he had attempted to sell tatty goods. This caution is the context for Drury's outburst about the cursed pins, blaming them and not Smeathman for the ragged wings of the Lepidoptera he sent from Africa.

Delivering perfect specimens was challenging work, and Abbot's success as a supplier of natural history specimens was due in part to his extraordinary skills as a taxidermist. However, even Abbot was scolded by Henry Escher about a shipment of insects that arrived in Switzerland somewhat the worse for wear. Abbot warned Swainson, who was about to send some Brazilian butterflies to Escher, that "he is very particular in his Directions in packing up the Insects to be sent to him." Apparently, the journey from Hamburg to Zurich, a great distance overland, resulted in Abbot's insect shipment receiving "great injury." Abbot reassured Escher that the box he had sent had been "securely packed up" and suggested that the damage must have been done by "ignorant persons . . . opening them at the Custom house."[84] Swainson would have appreciated this story not only as a warning about dealing with Escher but also because his father, T. J. Swainson, was collector of customs at Liverpool, the port through which Abbot usually sent his boxes filled with insects and drawings destined for London collectors. The elder Swainson, an amateur conchologist, was known to waive customs duty fees for fellow naturalists transporting their specimens, and Abbot could count on Liverpool's customs officers to carefully unpack his shipments and provide other forms of special treatment.[85] But good pins, strong boxes, and lots of camphor, plus cooperative customs officials, were not enough to ensure that specimens would arrive at an intended destination. As we shall see in the next section, forces beyond the control of specimen hunters could undermine their best efforts.

Difficult and Dangerous Work

Collecting specimens was difficult work and could also be quite dangerous. With over sixty years of experience in Georgia's woodlands, swamps, marshes, and creeks, Abbot must have had his share of adventures, if not dangerous encounters with wildlife. Late in life Abbot began his autobiography at a friend's prodding. He stated that he had had many adventures as entertaining as those found in novels, but he left off writing his autobiography after just six pages, and his adventures were left untold. Without Abbot's words, we are left to imagine what might have occurred: Copperheads lurking in the brush along blackwater creeks? Alligators in the swamps whose waters seeped in from the Ogeechee River as it flowed on its way to the Atlantic? Did he get lost in the woods, shoot himself by mistake, almost drown crossing a river, or become deathly ill from a bite or from drinking bad water? These scenarios may sound overly dramatic, but these incidents actually happened to other traveling artists and naturalists. For example, Abbot's contemporary Thomas Robins, a natural history illustrator, drowned in Jamaica trying to cross the White River, and the drawings he had done for his patron, John Ellis, a wealthy Jamaican planter, were lost along with Ellis when the ship on which he was returning to England went down at sea.[86] Audubon narrowly escaped an alligator attack during a birding expedition in the saltwater marshes along the Georgia and Florida coasts, and when Alexander Wilson crossed the Savannah River, his horse "threw himself into the torrent," and Wilson barely succeeded in rescuing it from drowning "at the risk of my own life." Wilson remarked how challenging Georgia's backcountry was to traverse: "I am utterly at a loss in my woods rambles here, for there are so many trees, plants, shrubs, and insects, that I know nothing of."[87] Whether Abbot suffered from injury we do not know, but he did become quite ill with fever shortly after arriving in Georgia, and when he was in his sixties, he became so ill that he had to delay by several months the delivery of drawings he had promised Escher, who had already paid for them.[88] Severe illnesses affected several famous traveling natural history artists. Sydney Parkinson, artist on Captain James Cook's first circumnavigation, died in 1771 from dysentery after stopping in Batavia (now Jakarta, Indonesia) on the return home, and Lewin, another naturalist artist in Drury's employ, almost died of the flux (dysentery) not long after landing in Australia. He was ill for six months and was very slow to recover his health, causing him to fall behind in his delivery of specimens to be sent to the impatient Drury.[89]

One of the most dramatic stories about the dangers of collecting natural history specimens was told by Abbot's agent John Francillon in a letter to John Leigh Philips, Francillon's friend and fellow insect collector. The story concerns David Burton, "a clever young man," according to Francillon, who was about to take up an administrative post in New South Wales. He was "a Gardener recommended by Mr Lee to Sir Joseph Banks"; his skills as a gardener may have appealed to Banks, who helped him to procure a position in the colonial bureaucracy. Before Burton left England for Australia, Francillon had persuaded this young man to collect Australian insects on his behalf. On April 7, 1792, less than a year after his arrival in the new colony, Burton accompanied a group of soldiers on a duck hunt, and it was during this expedition that Burton accidentally shot himself. The incident was described in detail by David Collins (1756–1810) in his history of the colony's early years:

> Reaching the banks of the river, they found on its surface innumerable flocks of those fowl of which they were in search. Mr. Burton, in order to have a better view of them, got upon the stump of a tree, and, resting his hand upon the muzzle of his piece, raised himself by its assistance as high as he was able. The butt of his piece rested on the ground, which was thickly covered with long grass, shrubs, and weeds. No one saw the danger of such a situation in time to prevent what followed. By some motion of this unfortunate young man the piece went off, and the contents, entering at his wrist, forced their way up between the two bones of his right arm, which were much shattered, to the elbow.[90]

The ensign in charge of the expedition stopped the bleeding with a tourniquet made from a torn shirt, keeping Burton alive for the journey back to Parramatta, which took a day and a half, during which Burton developed a fever. "In the night of the 12th the mortification came on, and he died the following morning, leaving behind him, what he universally enjoyed while living, the esteem and respect of all who knew him." Collins suggests that the new colony felt this loss greatly: "This young man, on account of the talents he possessed as a botanist, and the services which he was capable of rendering in the surveying line, could be but ill spared in this settlement."[91]

Francillon's narrative mirrors Collins's official one in the details of Burton's death, but he adds his own twist, revealing much about the dangers of being a specimen hunter and the disappointments that metropolitan collectors experienced as they worked to establish networks of exchange beyond their purview. Burton was ostensibly a superintendent of convicts, but he was also active as a botanist, a surveyor, and a collector of natural history specimens. Francillon wrote of Burton: "He undertook to collect Insects for me, and I agreed to purchase of him all that he could collect and send me. I furnish'd him with every article necessary for collecting, such as several large store Boxes Cork'd, many pocket D[itto] [Boxes], Netts, Pins &c &c and every thing in plenty that he could want. Before he went I took him out in the country several times, show'd him how to collect and expand their wings, which he did very well before he went." Not only had Francillon given Burton insect-collecting supplies, he had also trained him, teaching him how to capture butterflies and moths and to pin and preserve them. Burton sent a letter to Francillon saying he had arrived safely in New South Wales and had begun collecting for him, sending along with the letter a box of intriguing insects. Francillon told Philips:

> I receiv'd a Letter from him last Jan[uar]y with a small Box of Insects by Lieut Dawes, there were a few specimens of Coleoptera which I show'd you, and gave you 1 or 2 of each variety. In his letter he said he sent this small Box by way of sample to show me what beautifull insects there were in that part of the world. And by the *Atlantic* [ship's name] which they expected from China in about two months which would sail afterwards direct for England, by whom he should send me a very fine collection of Insects, in two or three Boxes.[92]

This sample of the wonders Burton had collected for him made Francillon eager for the return of the *Atlantic*, but when it arrived at Deptford, "I went on board and was inform'd no parcel came directed for me." He was informed that "poor Burton" was dead and "must have died soon after he sent me the small Box and letter." Francillon tells the same story as Collins: "In crossing a rough place, he rested the Butt end of his Gun upon the ground, and without considering the Danger, put his Hand upon the mouth of it, to help himself over the place. In that situation the Gun went off, and lodged its contents all up his Arm and Died of Mortification in Six Days." Here the story veers from Collins's narrative to reveal details about what happened after Burton's death:

> His Effects were all *sold by Auction* there, by order of the Governor, and the whole of his property sold for £36, amongst them was my collection, Boxes and all the implements, which were purchas'd by Mr. White[,] Surgeon there, and sent the three Boxes containing Insects to his Friend in London as a present, and it was very fortunate for me that this Gent[lema]n was a

> particular Friend of mine. When he show'd me the collection I knew my Boxes well, and I inform'd him of the circumstances, show'd him my Letter, and He very generously said I should have the first choice of the Duplicates a pair of each. For me, as a disappointment it has turned out very Fortunate.[93]

Francillon must have been pained seeing his boxes at a friend's house filled with insects collected by Burton for him, and although he makes the best of the situation by appearing to be satisfied with receiving the duplicate specimens, the fact that he told this tale to his friend Philips indicates that it must have been a bit of a shock and a disappointment. He claims that what he regrets is that he cannot give Philips any of the duplicates of these Australian specimens as he had planned to do: "I am sorry I am deprived of serving my Friends which would have given me great pleasure and satisfaction."[94] Of course, Francillon's telling of this story, in focusing on his loss of a source of exotic insects, trivializes the tragic death of this young man but conveys the intensity with which collectors such as Francillon pursued their passionate accumulation of exotic insects. It also reveals the dangers and uncertainties that field collectors faced when they traveled abroad.

Francillon's story about poor Burton also demonstrates how natural history collecting operated within the structures of colonial bureaucracy and commercial ventures. Knowing that Banks had nominated Burton for this position based on his skills as a gardener, not an administrator, Francillon must have realized that Banks was using the bureaucratic post as a way to insert Burton into a situation from which he could conduct botanical research and collect botanical specimens for Banks and Kew Gardens. Francillon saw this as an opportunity to enlist Burton's efforts as a field collector of insects, adding to Burton's extracurricular activities by training him and equipping him to gather, preserve, and ship insects back to England. Francillon's strategy of cultivating an agent or a "correspondent" in Britain's colonies was a way that natural history collectors acquired exotic specimens. Knowing residents in foreign locales and building up the reciprocal ties that bound such relationships were time- and resource-intensive strategies that could fail due to the hazardous conditions of field collecting, as we have seen with Burton. Francillon's equipping of a prospective colonial functionary was not unusual. As discussed in the previous chapter, Drury and other insect collectors pinned their hopes on travelers and sojourners, many of whom were, like Burton, part of the larger imperial project.[95]

Political strife between nations and within empires could also interrupt the flow of specimens to Britain and Europe from the American South. The American Revolution put a stop to Abbot's plans of shipping drawings and specimens to Drury in the late 1770s and early 1780s. In 1807 President Thomas Jefferson decreed an embargo on shipping with Britain that was not lifted for two years.

The War of 1812 with its restrictions on trade drove Abbot to exasperation, nearly putting an end to his business as a specimen hunter. Writing in despair over this war between the United States and Britain, he complained to George Ord, the Philadelphia ornithologist, about its negative impact on his work:

> At the commencement of the War I had undertaken to make a collection of stuffed Birds & as a complete collection of Drawings of them in colors as I was able for a Gentleman in England but last fall in despair of seeing peace restored, I retired into the Country after having made about 220 Drawings[,] throwed away a large collection of stuffed skins, have entirely laid it aside, & entered into another line of employment, where I am in hopes the mad and destructive Ambition of the rulers of the world can have but little interfere[nce].[96]

Sharing his frustration with Henry Escher over "the Embargo & War, & a concurrence of unlucky Circumstances, I have been such a Sufferer, that I had determined to quit Natl. history & retire to the Country."[97] However, Oemler and LeConte persuaded Abbot to collect for them during the embargo, with LeConte engaging him "to make a General Collection of & draw all the Coleoptera of Georgia for him, Then all the Lepidoptera, the Papilio, the Cat. Chrys. & Fly, without the plant, the Phalaena the Catr. & Moth, Male & female when different without the plant, & all the Aranea," a tall order that took Abbot years to fulfill.[98] By 1816, after the peace had been restored, Abbot had returned to his work of drawing and supplying specimens for British naturalists and was actively engaged in discussions with William Swainson about book projects and specimen exchanges. In addition to the wars between Britain and its American ex-colonies, the Napoleonic Wars made it very hard for French and other European collectors to acquire Abbot's specimens and drawings from his British agents. In 1802 Baron Walckenaer, the great French arachnid expert and taxonomist, had attempted to persuade Francillon, acting as Abbot's agent in Britain, to send Abbot's spider drawings to him in Paris, but the silversmith refused. He may have thought it was too dangerous and the drawings would be lost, or perhaps his patriotic feelings trumped international cooperation when it came to scientific work.[99]

On top of the uncertainty that politics brought to bear on the movement of specimens and drawings across borders and over oceans, the very nature of shipping was less than reliable, with at least two ships known to be carrying Abbot's drawings and specimens having gone down at sea in the early years of his sojourn in Virginia.[100] Much later in his career, Abbot experienced a serious setback when one of his shipments of drawings and specimens was destroyed in a Savannah docklands fire while awaiting the arrival of a ship meant to carry

Abbot's items, along with highly flammable cotton, to Liverpool.[101] In a letter to Swainson, Abbot described the devastation caused by the conflagration:

> Savannah 15th. Janry. 1820
>
> Dear Sir
>
> After writing my last Letter, I returned home and immediately packed my birds in 3 boxes, amounting to upward of 300 Specimens, together with my Insects in another containing upward of 1000, and brought them to Savannah, and unfortunately left them at my friend[']s Mr Oemler[']s till I could ship them to You, on the 8th—I was awaked out of my first sleep on the night of the 10th. By the alarm of a most terrible fire, which aided by a most violent wind, and other unfortunate circumstances, spread with irresistible fury and continued till it consumed the entire centre, and most opulent and trading part of the City—The loss of property is immense—My worthy friend Mr Oemler lost all his collections of Natural History, and what is worst nearly all his property—I lost my boxes, some Drawings &c, and another person, from whom I should have received the payment of a Sum of money, is also considerable a sufferer, that I have little or no hopes of ever receiving it—this will much hurt my present finances. This is the greatest calamity that I have witnessed and suffered by (except only a violent Hurricane, that continued its dreadful Ravages for 12 hours, since I [have] been in this country)—
>
> But Dear Sir, tho' much depressed in spirits, I suppose I must not give up to Despair—I will with a blessing go home, and begin again to collect—and hope yet to ship You a box of Birds some time in the month of March next—[102]

Abbot's frustration is palpable at having lost a year's worth of work of collecting and drawing specimens, but he ultimately took this loss in stride and began to collect again, continuing to do so into the mid-1830s.

Abbot's skills as an insect hunter and breeder also served him well as a natural history artist. To draw natural history illustrations that were meaningful for naturalists required artists to study the anatomy of their subjects. Most natural history artists, including Mark Catesby, Wilson, and Audubon, worked from dead specimens. Although they valued studying the living creatures in situ, which helped them when it came to depicting these creatures as if they were alive, natural history artists worked in studios surrounded by the dead specimens they were in the process of drawing. Amassing a collection of specimens was therefore essential to the business of being a natural history illustrator. Audubon and Wilson were famous for collecting their own specimens, Audubon killing thousands of birds as he set about the process of drawing them, and Wilson crisscrossing on foot huge swaths of the eastern half of the United States to shoot and collect birds.[103] Abbot's decision to live where he collected set him apart from other natural history artists who, after bagging their specimens, retreated from the field and returned to urban centers of natural history inquiry, where they produced their artwork. Abbot chose to live in rural Georgia at a remove from natural history centers, yet he managed to produce and distribute thousands of entomological watercolors to natural history collectors in Europe and the northeastern United States.

Alexander Wilson commended Abbot's willingness to live "in exile" to study and draw insects and birds. He praised him along with the botanists André Michaux and his son François André Michaux for their devotion to the task of depicting North America's natural world, "those worthy and indefatigable naturalists, who, impelled by an ardent love of science, became voluntary exiles from home and all its sweets, and subjected themselves to years of labour and peril, in personal efforts to examine and illustrate the natural history of this extensive Western empire."[104] Wilson was right to praise Abbot's dedication to his work as a naturalist artist, but he was wrong to suggest that Abbot regarded living in rural Georgia among the creatures he loved as a form of exile. On the contrary, Abbot may have experienced the few years he was living in Savannah in the early 1800s as a form of exile. In September 1813 he was eagerly awaiting his "move into the Country, about 54 miles from Savannah, in a healthful Situation, in a part of Georgia, I have not yet explored this not being settled till lately."[105] He was hopeful that this move would facilitate a return to field collecting: "I shall be extremely glad if times wou'd once again permit, as formerly, to devote my whole time, & Study in Drawing and Collecting."[106] In December 1813 he informed Escher that he was now living in Screven County and happily at work: "The middle of Octr. I removed from Savannah to the Country about 60 miles from it. It is a very desirable Situation to collect Insects, having a large River Swamp on one side of me, and the open Pine woods uncultivated on the other, Intermixed with fields. I propose to be very industrious next Year in collecting."[107] Abbot was quite industrious, collecting over eight thousand insect specimens for Escher. It did, however, take him two decades to supply Escher with these specimens, for, as he explained, "to make a good collection is the work of much time."[108]

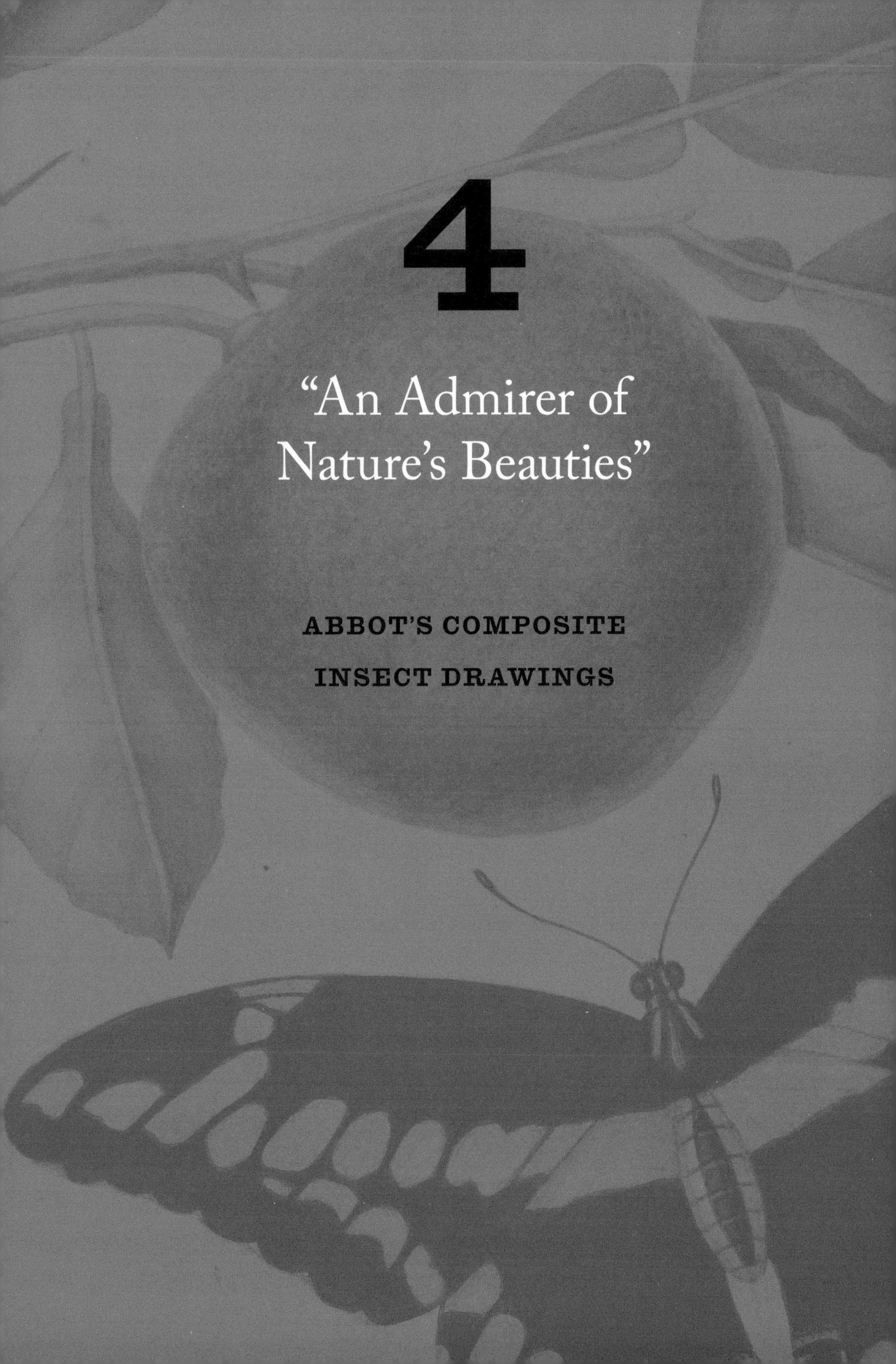

4

"An Admirer of Nature's Beauties"

ABBOT'S COMPOSITE INSECT DRAWINGS

FIGURE 4.1.
Abbot, "Catalpa Sphinx" (*Ceratomia catalpae*), watercolor drawing, plate 41, "Moths of Georgia" (ca. 1817–1820), Hargrett Rare Book and Manuscript Library, University of Georgia Libraries.

TWO ALBUMS OF ABBOT'S DRAWINGS held in the Hargrett Library at the University of Georgia consist of 105 watercolor drawings of mostly moths with their caterpillars, pupae, and host plants.[1] A properly credentialed visitor can view these drawings after being informed of the library's rules on the handling of rare books and manuscripts, rules that prohibit the use of pens, the wearing of jackets and carrying of umbrellas, the chewing of gum and sucking of cough drops. Sitting at a table in the reading room, the visitor awaits the arrival of the drawings held in the special collections vault. The drawings arrive encased in a specially designed archival box that cradles the two volumes in foam, protecting them from damage. The drawings are bound with what looks to be twentieth-century high-end binding with leather boards and marbled endpapers with embossed lettering on the spine that reads "Moths of Georgia."

The first time I opened one of these quietly elegant volumes, I was surprised and somewhat dismayed to be greeted by the previous owner's name, T. McDougal, on the first page in blue ink, most likely from a mid-twentieth-century ballpoint pen. The quiet gasp that escaped my lips upon seeing the inked signature stemmed in part from my training as a scholar who conducts research in archives and rare book libraries where pens are forbidden for fear their use will deface (usually accidentally) valuable prints, drawings, and books. In addition to the inked signature, each drawing was marked with annotations written in pencil identifying the plants and sometimes the insects. I regarded the penned signature and the many penciled notations on Abbot's drawings as having marred the beauty of rare artwork. However, these annotations indicate that the previous owner had valued these drawings for what they could tell him about the natural world, perhaps helping him to identify some plants that may have been growing in his garden. I do not claim that my pen-free approach to these drawings is the "correct" way to encounter these drawings, although the librarians might have been alarmed if I had wielded a pen as I examined them. The previous owner was clearly engaged with these drawings, deriving knowledge and pleasure from them, and he was arguably using them in the way they were initially intended to be used when they were drawn in the early nineteenth century.

a Sphinx
Catalpa
Bignonivides
Grows in the Swamp
near Savannah river
41

20

+++79.

+++No. 79. Libellula.

A Male of No. 80.

Taken 17th March & 15th July, about branches in Oak Woods.

The consternation I felt at the sight of the blue-ink signature and the many penciled notations on these precious drawings was not all that different from how I reacted when I first saw the red-inked stamp of a crown accompanied by the initials of the British Museum that appear on every one of Abbot's three thousand drawings held in the Zoological Library of London's Natural History Museum. This stamp declares the British Museum's and ultimately the Crown's proprietorship over these drawings, claiming them for the nation as part of its cultural and scientific patrimony. The red-inked crowns also make these drawings less attractive to would-be thieves tempted to surreptitiously slice out a few choice drawings to sell on to others wishing to own rare natural history illustrations. Libraries with valuable drawings and prints, including the British Library and the Library of Congress, have been victims of this kind of thievery, with visitors slipping drawings or maps in between the pages of their notebooks or in the lining of their trench coats. Nowadays, libraries are safer from patrons' light fingers with restrictions in place that prohibit patrons from bringing certain items (notebooks, coats, briefcases, and the like) into the reading rooms and with the surveillance of guards who roam up and down the aisles and the installation of video cameras.

FIGURE 4.2.
Abbot, "Libellula," male common whitetail (*Plathemis lydia*), det. J. K. Wadleigh, watercolor drawing, vol. 12, Francillon Collection, Archives and Library, Trustees of the Natural History Museum, London.

Although I silently groan when I see that stamp, I begrudgingly acknowledge that it has done its job protecting remarkable drawings from disappearing into the sometimes unethical trade in antiquarian books and natural history illustrations.[2] In addition to the stamp that rests on each of the museum's drawings, there is often some writing in pencil beneath the insect, an annotation perhaps by one of the museum's curators in his attempt to identify the drawn insect. On some of the other drawings, the penciled labels have been crossed out, and different names of the genus and species have been written above the rejected name. These are the traces of arguments that entomologists have had while trying to identify the insects in Abbot's drawings, arguments that may go back to 1818, when the British Museum first acquired these drawings. While I may moan and groan about these marks in pencil and ink on what I regard as *my* beloved drawings, I also have to admit that I am fond of these marks, because they are clues to how other people who came before me used and felt about these drawings. Over the years of studying Abbot's artwork, I have had to learn to share his drawings with these folks from the past (and future) and to acknowledge the validity of their engagement with these drawings: the penciled line that runs through the name of one species is a sign of passionate intellectual engagement, while the blue-ink signature on the first page of the Hargrett's volume tells us that these drawings were loved and a source of pride.

I call attention to the marks left by others when encountering these drawings because I want to stress that these (and all) drawings possess a materiality that cannot be reduced to the image that appears on the sheet of paper. A photograph of one of these drawings cannot convey what I experience when I see the actual drawing. Any rare book library aficionado will tell you that a two-hundred-year-old book that has sat in a special collections vault carries with it the lovely musty smell of old leather, a smell that precedes the act of opening the book. There is also the sound that the paper makes as pages are turned. The paper used by Abbot for the Hargrett's moth drawings has a substantial heft to it and a stiffness that does not interfere with its ability to bend where it should.[3] The watercolor paints Abbot used were of such high quality that they have retained their luminosity and saturation to this day. He made his own watercolors, as did most artists in this period, probably ordering pigments and gum arabic from Savannah's apothecary shops.[4] In this chapter, I pay attention to the material quality of Abbot's drawings while aiming to capture their lives as objects as they traveled over space and through time.[5]

Hargrett's album of drawings with their graceful depictions of the lepidopteran life cycle—moth, caterpillar, chrysalid, and host plant—is not unique. A set of drawings, for instance, duplicating many of those in the Hargrett Library is in the National Library of New Zealand, while drawings similar in style, format, and subject matter are to be found in London's Natural History Museum, Harvard University's Houghton Library, and Zürich's Zentralbibliothek. The drawings in New Zealand have been described by art historian Janice Neri as having "Abbot's characteristic style": "The drawings depict in soft, glowing tones butterflies, moths, beetles, flowers, and other plants arranged in delicate and elegant compositions."[6] Drawings with this arrangement of multiple insect bodies and food plants, however, represent only a small portion of the thousands of drawings Abbot produced. I have chosen the label "composite" to distinguish these compositions of multiple insects from the majority of his drawings, which display a single insect on a small piece of paper, the topic of the next chapter. This chapter deals with the larger drawings in quarto that feature multiple figures arranged in a decorative, geometric, or linear layout.

Over six thousand of Abbot's entomological drawings are extant, and they have their own stories to tell about how they came into being, how they were used, where they are now, and how some of them were valued greatly while others were disregarded and abandoned. Most of Abbot's drawings were initially purchased as objects of aesthetic and scientific interest; some were given as gifts within natural history circles; and some were sold at auction and entered the marketplace after owners died and their heirs disburdened themselves of unwanted items. In tracing some of the trajectories of Abbot's drawings once they left his hands in Georgia, this chapter and the next examine how collectors acquired and used Abbot's drawings. I take the marks in pencil and ink on Abbot's drawings as my entry into the role that drawings played as part of the vital material and visual cultures of natural history collecting, display, and classification of that era. The annotations, inscriptions, and transcriptions on Abbot's entomological drawings are the visible remnants of complex forms of engagement performed by natural history collectors, amateur naturalists, natural history authors, and professional entomologists. Before turning to what these marks in pencil and pen can reveal about what these drawings might have meant to those who wrote on them, we need to consider the materials that Abbot used to create these lovely objects. Much of their appeal is due to his skillful handling of his medium. I take this moment to direct our attention to a key component of his drawings: the paint he used to convey the colorful bodies of arthropods.

Making Watercolor Drawings

To appreciate fully the materiality of Abbot's drawings, we need to inquire into the processes that enabled him to produce thousands of such highly valued watercolor drawings of insects and spiders. What he achieved is remarkable, especially given that he lived mostly in rural Georgia, working far from urban centers of natural history inquiry and their infrastructures that supported natural history illustration. Aside from his talent and skill, the quality of the images that he produced was dependent in part on the quality of the materials he used. To be able to paint a tiny hair on a fly's body or the intricate pattern on a spider's abdomen, morphological characteristics entomologists and collectors wanted to see, he needed materials to behave properly. He required fine paper that was absorbent, but not too absorbent, and flexible enough to bend, yet sturdy enough to tolerate the application of color washes. He also needed paint that was vibrant and did not fade; watercolor brushes capable of fine detail; graphite pencils that were not hard and gritty; and gum arabic, a key ingredient in watercolor paint.[7] Where did he acquire these art supplies?

Had he been in London, he would have had easy access to an array of shops that sold such art supplies. Fine paper, manufactured especially for watercolor paint, could be bought in stationery stores such as Dorothy Mercier's shop "at the Golden Ball in Windmill Street," where "might be had all Sorts of Paper for Drawing, &c.," including "English, Dutch, & French Drawing Paper, Abortive Vellum for Drawing," "Camels hair" brushes, and "the best Black Lead Pencils."[8] Color shops, specializing in selling pigments, most of which were derived from ground minerals and chemically manufactured compounds, offered their customers, who were a mix of professional and amateur artists, pigments such as white lake, carmine, vermilion, and verdigris. Pigments mixed with linseed oil and squeezed into clean pigs' bladders were widely available in London, as they were also sold in apothecary shops, sometimes referred to as druggist, chemist, and oil men's shops.[9]

Options for buying pigments in Georgia were limited during the period when Abbot was the most productive, which was between 1790 and 1820. He may have acquired art supplies directly from England in the late 1780s and

1790s. Drury may have shipped to him pigments, paper, and other materials during this time, since we know that during the 1790s Drury sent paper, pigments, black ink, and copper plates for engraving to John William Lewin, who was living in Australia near Sydney. It is also possible that Abbot may have ordered pigments and other art supplies from Philadelphia shops that dealt in imported luxury items. For instance, on November 27, 1776, Robert Bass, apothecary, advertised in the *Pennsylvania Gazette* that he was selling "just imported" goods from London on Market Street in Philadelphia: olive oil, Italian prunes, "Drugs and Medicines of most Kinds," and "Painters Colours, Varnish, Lacker, &c.&c.&c."[10] During the colonial and federal periods, apothecaries were key players in the color trade. They imported from Britain chemicals such as arsenic, sulfur, mercury, and magnesium, which were used to make elixirs, ointments, and tinctures. Minerals such as cinnabar and ocher and the synthetic chemical compounds such as vermilion and Prussian blue that were used to make painters' colors were part of this larger trade in chemical products, which also included dyes for fabric. The Revolutionary War put a stop to this trade, but soon after the peace had been reestablished between Britain and the United States, apothecary shops in Philadelphia did a brisk business selling British-made painters' colors along with other chemical elements and compounds. In 1786 Robert Bass was supplying his customers with "Reeves' new invented, incomparable, Drawing Cakes, for miniature painting," referring to dried cakes of watercolor paint manufactured by William and Thomas Reeves in London. In 1788 Bass's advertisement stressed the popularity of Reeves' watercolors for sale at this shop: "assorted boxes of the famed Reeves' patent colours, so singularly distinguished for their excellency in miniature drawing, and hold their fine teints [tints] and shades."[11] Savannah and Charleston lagged behind Philadelphia in the color trade, but by 1809 Savannah's newspapers were advertising the sale of imported pigments. Alexander Hunter announced in the *Savannah Evening Ledger* that he was selling on Harden's Wharf a newly arrived cargo of painting supplies:

> English white Lead, ground in oil and dry
> Yellow Ochre, Spanish Brown in oil and dry
> Venetian Red, ground in oil and dry
> Boiled Linseed Oil in jugs
> Japan and Copal Varnish, Whiting, Allum
> Green Copperas, Umber, Litherage [Litharge], Red Lead
> Patent Yellow, Ivory Black, Black Paints
> Lampblack, Vermillion, Rose Pink[12]

By the 1820s shops in Savannah, some owned by apothecaries, were advertising impressively long lists of pigments. Anson Parsons, a "druggist," sold painters' pigments among a range of chemical products out of his shop, which he called the Perfumery. Hazard and Denslow, purveyors of paints, sold a "general assortment of Paints and Painter's articles," which included yellow ocher, Chinese vermilion, red lead, white lead, flake white, and lampblack, along with camel-hair brushes and gum arabic.[13] By 1823 "elegant boxes" of Reeves' watercolors in that familiar lozenge shape were sold packed in specially designed wooden boxes with room for brushes, palettes, and other watercolor tools.[14] Lay and Hendrickson's advertisement for "Reeves' Water Colours in Mahogany boxes, in great variety, Reeves' water colours, for children" suggests that these expensive items were for genteel amateur artistic pursuits, more in keeping with notions of upper-class cultural refinement than for working artisans and professional artists.[15]

As a professional artist, Abbot would not have bought Reeves' ready-made cakes of dried watercolor paint. They would have been too expensive, and he would have preferred to have control over making his own paint, which was his chosen medium. He would have bought pigments, sometimes known as "London Ground Colours," from shops such as those mentioned above, but these pigments were not yet suitable for mixing with water to make paint.[16] Most commercial pigments were the product of lengthy chemical processes, some quite complicated and dangerous. White lead, for instance, involved arsenic, and vermilion was manufactured by heating a combination of sulfur and mercury to evaporation rates and then collecting the residue along the inside of the heating chamber. Much had to be done to transform these chemical elements and inorganic chemical compounds into proper translucent watercolor paint. These pigments had to be soluble in water and able to adhere to paper, and they had to be sufficiently stable to retain their colors over time.[17]

For instructions on how to transform commercial pigments into watercolor paint, painters could refer to the many trade and art manuals published during the eighteenth century. One such manual, *The Art of Drawing, Painting in Water-Colours* (1757), offered painters advice on how to render pigments such as red lead "fine and fit for Use, for, as it is bought at the shops, it is not by any means fine enough to paint with."[18] In Georgia Abbot happened upon a copy of the 1757 edition of *The Art of Drawing, Painting in Water-Colours* that promised to provide "Instructions for preparing, mixing, and managing all Sorts of Water-Colours used in Painting, so as to represent Nature in the greatest Perfection."[19] Remarking on this manual, Abbot wrote in the introduction to the manuscript of *Insects in Georgia*: "I took it up with much disdain, not thinking it cou'd contain any thing worthy of Notice after my having Read many pompous Works, joined to my own vast Experience."[20] Unfortunately, Abbot's manuscript stops at the bottom of the page, the rest having been discarded, leaving this fragment, "But I think it without," which suggests he was about to say "without a doubt" and then something about how valuable this manual was. We can only guess what recipes he might have found useful. Fortunately, copies of the multiple editions of this book along with many other manuals are extant and can help us to imagine the processes by which commercial pigments could be transformed into watercolor paint.

Watercolors, unlike oil paints, are water-based solutions of finely ground organic and inorganic pigments. To make pigments soluble in water (if they were a mineral or naturally occurring inorganic chemical compound such as ochre or a manufactured chemical compound such as vermilion), they would have to be further ground down into a fine powder. To achieve this, an artist would have had to grind these mineral-based pigments against a stone or ceramic slab using a muller. This required strength and skill. To make, for instance, red paint from "Lake," which consists of "the grosser Particles or the Lees of Carmine[,] requires a great deal of Grinding" to create a powder fine enough to be evenly dispersed throughout the water.[21] To keep the pigment particles suspended in water and to make the paint adhesive, the artist first added powdered gum arabic to the water, making what was called gum water. Gum arabic is made from the dried sap of the *Acacia senegal* tree, and the amount added to the water depended on the chemical properties of the pigment. Acting as an emulsifier and a binder, gum arabic was (and still is) a key ingredient in watercolor paint. Imported from Africa, it circulated widely within European commercial circuits, in particular within the medicinal and chemical trades, appearing, for instance, in Hazard and Denslow's Savannah shop.[22]

Along with advice on how to turn commercially available pigments into watercolor paints of all different colors, recipes for preparing gum water can also be found in many eighteenth-century art and trade manuals. *The Artists Assistant* (1794) published this recipe: "Take some gum-arabic of the whitest sort, bruise it, and tie it up in a piece of woolen cloth, and steep it in spring-water in a glass or earthen vessel, till it be dissolved. If it be too stiff, add more water; and if too thin, [add] more gum. With this water you may temper most of your colours; using such a quantity, that being touched when dry, the colour will not come off. If the colour shine, there is too much gum in it."[23] The right amount of gum arabic could keep paint from being too runny, but too much could make the dried watercolor painting too shiny. Some artists applied gum water directly to the paper to prepare it to receive the watercolor paint. Again, too much or too little gum arabic in the solution could cause problems. Offering novice watercolorists advice on preparing their paper for the application of watercolor paint, George Edwards, a prolific natural history illustrator and an acquaintance of Abbot's in London, cautioned that the paper "ought to be neither over nor under-gummed: that which is too much gummed, or seized in the making, is so hard and close it will not take in the colours at all. . . . An under-gummed paper hath a contrary inconveniency; for the colours are apt to run through it, and spread beyond your design on the outline." His solution to this problem of overgumming or undergumming one's paper was to touch the paper "with your tongue." He explains

FIGURE 4.3.
Abbot, cecropia moth (*Hyalophora cecropia*), watercolor drawing, "Insects of Georgia" (1827), Stuart A. Rose Manuscript, Archives, and Rare Book Library, Emory University.

that "an ungummed paper will stick very strongly to the tip of the tongue when touched: an over-gummed paper will hardly stick at all; by which a proper medium may be found, that only sticks a little to the tongue."[24] Edwards's somewhat vulgar but simple solution to the problems of adhesion that watercolorists faced when applying watercolor paint to paper underscores the kinds of embodied artisanal knowledge that artists needed to possess in order to do their work.

We have no way of knowing if Abbot used his tongue to test his paper's ability to absorb paint, but he may have, since he had met Edwards when he was just starting out as an artist. Edwards, who had encouraged Abbot with praise for his drawings, may have shared this tidbit of advice. A supplemental ingredient Abbot may have added to his gum water was coloquintida, a purgative made from the unripe fruit of the colocynth, commonly known as the bitter cucumber. Several manuals encouraged watercolorists to add coloquintida to their paint. Why? "To keep the flies from your work." *The Artists Assistant* suggests: "When you have made your gum-water according to either of the foregoing directions, add thereto a little coloquintida, which will preserve your work, when exposed, from being spoiled by the flies."[25] Why, you might ask, would flies be eating the watercolor paint off the paper and spoiling an artist's work? It turns out that in addition to gum arabic, sugar or honey was added to the paint solution to enhance the paint's plasticity. Without sugar, as the watercolor mixture of pigment and gum arabic dries on the paper, it can become brittle and flake off the paper. Mixing a bit of sugar with the gum arabic solution allows the dried watercolors to adhere and bend with the paper, necessitating the addition of coloquintida as a fly repellent.

Recipes for making watercolor paints were something Abbot took seriously. In the early 1790s, when he was working on *Insects of Georgia*, he complained that he had had trouble with the quality of some of the watercolor paint he had made. Frustrated by the failure of watercolors to convey the intensity of the colors of the insects he drew, he admits to having the same failing as Eleazar Albin (active 1713–59), a natural history illustrator who, according to Abbot, "in his Hist. of Insects says he has not painted them of too bright colours, but poor Albin like myself he falls much short of the Originals for want of sufficient bright colours." Abbot confessed that he had had trouble with his greens: "In some of my first Drawings the Greens are turned blackish, owing to my then using Sap Green, I now quite discard it, & use a mixture of Gum Bouge & Indigo, but I even find that this fades some."[26] Sap green, gamboge, and indigo are plant-based pigments (organic chemical compounds); these pigments were quite unlike vermillion or white lead, products of complex chemical processes conducted by commercial manufacturers. A recipe for sap green that appears in *The Art of Drawing, and Painting in Water-Colours*, the book Abbot referenced, describes how to make this less than stable pigment: "Take the Flower of the blue *Flag-Iris*, or *Flower-de-Luce*, and press them while there is any Juice to be got from them; boil this gently in a glazed Pipkin, till it grows thick, adding a little Allum to it, and it will make a very useful and lasting Green." This color is "like that of an Oak Leaf, if it is used thin with common Water" (as opposed to using gum water), "but it will, if we use it strong, produce as dark a Green as any we can imagine."[27] Abbot seems to have disagreed with the manual's assertion that sap green made a "lasting Green." Dissatisfied with the instability of sap green, he hoped that the green he made from gamboge and indigo would be longer lasting. He may have made his own gamboge, a yellow pigment produced by boiling "French Berries" in a "Pint of Lixivium made of Pearl-ashes and Water," a recipe found in *The Art of Drawing*.[28] This manual may have also given him the idea of mixing indigo with the gamboge. French berries could be bought in Savannah at Lay and Hendrickson.[29] Indigo, a dye originally produced in South Asia from the plant *Indigofera tinctoria*, was valued for the intensity and depth of its blues. The demand for indigo outstripped its supply until it was cultivated in the mid-eighteenth century by an enslaved workforce on plantations in Spanish Central America, the French Caribbean (Haiti in particular), and South Carolina. Indigo was also for sale in Savannah at Hazard and Denslow's shop.[30] As a natural history illustrator, Abbot was deeply concerned about his colors retaining their vibrancy and mirroring the natural colors of the insects he drew.[31]

Abbot's clients expected his drawings to be accurate representations of the creatures he drew. Abbot would have learned from Drury how important color was to the process of identifying and classifying insects.[32] He would have seen firsthand how much Drury cared about the accuracy of the illustrations in his multivolume opus, *Exotic Insects* (1770–82). He would have witnessed or at least heard about Drury's frustration with his illustrator, Moses Harris, over the quality of the illustrations' coloring. Drury superintended Harris's coloring of the book's engraved illustrations and complained when they did not measure up to his standard. Writing to Harris, Drury aired his irritation over poorly colored illustrations:

> Those 3 sets you did last you have made ye [the] spots in each of ye underwings or rather ye patches that are a beautiful saxon green in ye fly in ye plates are a mazarine blue & that part that runs over ye scarlet eyes on ye abdominal edges, you have made a pea green.... Likewise in 2 other setts this figure is colod. [colored] blue in one wing & green in ye other.... I fancy you did them by candle light or you could not have erred so much.[33]

Drury insisted on accurately colored plates; he announced in the preface that his books contained "*just* and accurate figures": "The utmost care and nicety has been observed, both in the outlines, and engraving. Nothing has been strained, or carried beyond the bounds nature has set; and whoever will compare the engravings with the originals, I flatter myself will allow, that nothing is borrowed from fancy; or any colour given to an insect that does not really exist in the subject intended to be represented."[34] To identify insects, collectors and entomologists needed to take color into account. However, color's importance went beyond its use in identification, having to do as well with an aesthetic appreciation of the colorful patterns that insect bodies and wings display.

The remarkable variety of colors of insect bodies captivated Eleazar Albin, an artist and art teacher who produced an illustrated book, *A Natural History of English Insects* (1724), which Abbot owned and referred to several times in his correspondence. Albin drew, engraved, and colored the illustrations himself, explaining in the preface his reasons for doing so: "Teaching to Draw, and Paint in Water-Colours being my Profession, first led me to the observing of *Flowers* and *Insects*, with whose various Forms and beautiful Colours I was very much delighted, especially of the latter several of which I painted after the Life, for my own Pleasure."[35] In short, the color of insects was an integral part of the experience of collecting insects. Abbot, raised in this culture of *ad vivum*—truth to life—was scrupulous about capturing the colors of his chosen subjects, making his own paint and applying it with care, and determined to fulfill Drury's admonition to paint "just and accurate figures."

Abbot's Watercolor Technique

Abbot drew a set of one hundred drawings, all bound together in an album, in the late 1760s and early 1770s when he lived in London and was under Drury's mentorship. This album containing Abbot's earliest work was once owned by Andrey Avinoff, who, as a lepidopterist and an artist, was well positioned to evaluate and explain why he thought these drawings were remarkable. He valued them not only for Abbot's accurate rendering of insect bodies but also for his deft handling of the medium, Avinoff praising Abbot's ability to render "with microscopic perfection" the details in a beetle's spiny legs or a fly's hairy body: "Every hook and every spine is recorded with astounding precision." Abbot's attention to detail, however, does not override the "general effect," as Avinoff explains: "The luster of the elytra and the transparency of wings is never lost by an excess of details. The venation of the smallest flies is never exaggerated in emphasis to the point of losing the general gauzy appearance." Of Abbot's drawing of a death's-head hawkmoth, Avinoff wrote: "One peculiar pattern of the front wings of this moth is produced in nature by light scales of different shades and density on a dark background. The same effect is rendered by Abbot without the use of white paint and producing to the naked eye the absolutely accurate and true aspect. A magnification of 7–8 times discloses a most ingenious, uniform and astonishing technique of microscopic strokes of dark tints on white paper showing through."[36] This exquisitely detailed rendering of insects' bodies, in addition to being aesthetically pleasing, is central to conveying scientifically accurate morphological information about those bodies.

Because Avinoff also drew butterflies and flowers, he could appreciate Abbot's skillful application of watercolor, praising his method as "'pure' watercolor technique, using transparent tints without washes of gouache." What Avinoff meant by "pure watercolor technique" can be inferred from what he wrote about his own methods when coloring his drawings of plants: "Only transparent pigment, no use of opaque colors, no white anywhere, not a single stroke in highlights in the whole set (although it would often simplify the execution). All light parts are lighter washes of pigments. I preserved through[out] 'thoroughbred aquarelle.'" Using the paper to shine through lightly applied translucent paint is for Avinoff the sign of an expert watercolorist, someone who does not use white and opaque paints and instead plans ahead, holding back in the application of paint as a way to create light in the image. Referring to Abbot's drawings of insect wings and their venation, Avinoff states: "An artist versed in the technique of watercolor will appreciate the difficulty of preserving the fine light veins on a colored background without using white paint."[37] Avinoff saw in Abbot a fellow "pure" watercolorist, and perhaps he treasured Abbot's drawings as an exemplar of what he himself worked hard to achieve in his own artwork.

These particular drawings were much prized by Avinoff, who proved to be an excellent caretaker of these precious drawings, which are among Abbot's earliest extant, dating from 1767 to 1773. First, Avinoff did not abandon them to the flames of the Russian Revolution (his library in Ukraine was burned), and second, when he retired from the directorship of the Carnegie Museum of

Natural History in the mid-1940s, he made sure they found a permanent home there along with the other precious entomological books he had collected over the years to compensate for those hundreds lost during the Russian Revolution. He wrote the following about these drawings in a letter about trying to assess the value of the rare books he was about to bestow on the library of the Carnegie Museum of Natural History.

> It is the *only* book I took with me in a suit case when I left Russia for good, since I considered it the most precious piece in my old library. Really, it is not without emotion that I part with it, since it is a set [of] *unpublished* originals of the most outstanding American entomologist of the XVIII Century. Some day it *should* be reproduced and a descriptive work on it should have been written long ago. But it would require considerable research to identify the species. I know, incidentally, that it contains the figure of Terias (Eurema) adamsi; described by my friend Lathy sometime about 1910, and here it was figured and remained unknown for 145 years![38]

Clearly, Avinoff loved these drawings. For him, they represented the perfect union of art and science: "Here the artist was in full control of the purposes of the naturalist." Managing to convey enough details about the insect's anatomy to aid in its identification, Abbot had at the same time attended to the aesthetic qualities of arthropods' bodies, the wasp's gauzy wings, the luster of a beetle's back, and the dappled colors of butterflies' wings.[39] The intensity of Avinoff's attachment to Abbot's drawings and the other rare entomological books in his collection is revealed in a 1945 letter he wrote about his impending gift: "I feel happy that the Carnegie Museum, which treated me so grandly throughout all the years I served this institution, will become the repository of my treasured possessions."[40]

Avinoff's collection of Abbot's drawings is also quite remarkable because Abbot's annotations appear on the back of almost all of these drawings. His annotations note the date each drawing was made, the common names of the insects if known, and to whose collection the specimen belonged. For instance, written on the back of a drawing displaying four little yellow moths in a pattern with four brown moths are the words "J Abbot ad vivum delin. 15 Aug: 1772," which means that he drew these specimens from life rather than copying them from an illustration. Beneath this statement, he reveals that two of these moths were "from America [collected] by Mr Kuckahn," one of them "in Mr Drury's Cabinet" and the other "in my own cabinet."[41] Kuckahn, who lived in the colony of New York during the 1760s and 1770s, supplied Drury with hundreds of specimens, and apparently Abbot was given some of Drury's or he was able to purchase moths from this supplier. The frequent mention of Drury's cabinet in Abbot's annotations suggests that Drury happily gave Abbot access to his collection and may have also facilitated access to other cabinets belonging to prominent collectors such as Margaret Cavendish Bentinck, Duchess of Portland. Her name appears next to the name of the "Marvel du jour" moth, both written by Abbot on the back of his drawing of several brown moths.[42]

The back of another one of his moth drawings has the following inscription in Abbot's hand:

> John Abbot ad vivum delin. 17th April 1772
> 282. Male Frosted Orange came out 14 Sep:
> 284. Female Frosted Orange came out 23 Sep: Gortyna? [in another hand]
> 286. Male Bramble taken on Norwood 25 Oct on a Blackberry
> 288. Female[43]

These notes tell us that the bramble shoot moth (figure 286) was caught by Abbot as it was sitting on a blackberry bush in Norwood in October, and the frosted orange moth (figure 284) emerged from its chrysalis in September, which suggests that Abbot had raised it, probably catching a caterpillar that underwent metamorphosis under Abbot's watch. These notes testify to Abbot's skills as a breeder and a hunter of lepidopterans well before he arrived in the American South.

On the back of a lovely drawing of some of England's prettiest butterflies, Abbot has written the date he completed the drawing, January 30, 1773, and identifies the butterflies as a "Purple Emperor" (male and female) and a "Camberwell Beauty." Intriguingly, Abbot's annotations have been copied in black ink onto pieces of tracing paper that have been tipped into this album. Tipping a sheet of paper into a volume implies that this volume was bound or rebound to include several of these traced annotations, something that was done before the album was purchased in London by a Russian bookseller for Avinoff in 1913. Next to the notations in ink are penciled words written in elegant handwriting that is not Abbot's, words that correctly identify the purple emperor as *Apatura iris* and the Camberwell beauty (known in the United States as mourning cloak) as *Vanessa antiopa* (now *Nymphalis antiopa*). The penciled IDs are an example of how Abbot's drawings became sites of inquiry and exchange of ideas concerning identification and classification, which, in this case, consists of adding to the proper Latinate Linnaean name for these species.

Abbot's Early (and Late) Style: Decorative Configurations

The drawings that Avinoff owned and gave to the Carnegie Museum were painted when Abbot was quite young (sixteen to twenty-two years old) and just starting out as a natural history artist. These drawings might have served as samples of his work, functioning as a portfolio to display his virtuosity as an entomological illustrator and to demonstrate that he was an artist with a naturalist's eye. The composition of these drawings is geometrical, with several insect bodies displayed in a pattern on a page. Another album of drawings that Abbot made while still in England, carrying the date of 1772, shares with the Avinoff volume a highly stylized display of insect bodies; however, instead of geometrical patterns, the insects in the 1772 volume are laid out in "neat rows" in a tabular format.[44] Both formats mimic the kinds of arrangements of natural history objects that were found in the glass-covered drawers of seventeenth- and early eighteenth-century cabinets of curiosity. Such designs refer to an older way of presenting natural history specimens, prior to the introduction of Linnaean systematics and when aesthetics primarily determined the arrangement of specimens (figure 4.4).

With the exception of two spider albums (to be discussed in the final chapter) created in the 1790s, Abbot did not use the geometrical, patterned format for his entomological illustrations again until the late 1820s. When he was in his seventies, he made albums of drawings laid out in decorative configurations much like the Avinoff volume. In these late drawings, the insects are painted in an even more stylized way, with very flat but colorful bodies rendered nearly abstract in their patterned arrangement. Three such albums, dated 1827, 1828, and 1830, are now held respectively in the Rose Library at Emory University, the Kenan Research Center of the Atlanta History Center, and the Houghton Library at Harvard University. Emory's bound volume of Abbot's drawings, dated 1827, has eighty-five drawings on wove paper with the watermark Butler, an American paper mill. Entitled "Insects of Georgia, consisting of the most remarkable and beautiful kinds drawn and coloured from Nature by John Abbot," it was originally owned by Godfrey Barnsley (1805–73) of Savannah. The 1828 volume, with one hundred watercolor drawings, held at the Kenan Research Center, has the same title as Emory's and was owned by Samuel Wright (1793–1841), a business associate of Barnsley's (see figures 4.5 and 4.6).[45]

The Houghton Library's 1830 volume was a gift to Thaddeus William Harris, a Harvard librarian and an entomologist, not from Abbot, though they had been correspondents, but from Edward Doubleday, Harris's British friend who eventually was employed in the British Museum's Lepidoptera Department. In 1839 Doubleday, who knew of Harris's interest in Abbot's work, wrote to Harris to say he had spotted lepidopteran specimens that Abbot had collected in the British Museum. He also explained that he had found in a London bookshop an album of Abbot's drawings that had been owned by Swainson. Doubleday bought it "as a trifling present" for Harris because this "small folio volume," containing "one hundred and fifty figures of Georgian Coleoptera, and about three hundred and fifty of Lepidoptera," might have drawings of species that would be new to Harris, as they were to Doubleday.[46] Harris gave these drawings his close attention, numbering all the Lepidoptera and writing in pencil the scientific names of many of the moths (see figure 4.7).

FIGURE 4.4.
Abbot, "A Natural History of Insects" (1772), watercolor drawing, plate 13, MS Typ 426.1, Houghton Library, Harvard University.

FIGURE 4.5. Abbot, two emerald euphorias (*Euphoria fulgida*), (female) eastern Hercules beetle (*Dynastes tityus*), two common green June beetles (*Cotinus nitida*), det. J. K. Wadleigh, watercolor drawing, "Insects of Georgia" (1827), Stuart A. Rose Manuscript, Archives, and Rare Book Library, Emory University.

After Harris's death, the Boston Society of Natural History purchased this volume from his estate. In the 1880s Samuel H. Scudder (1837–1911), one of the leading American lepidopterists of his time, wrote quite disparagingly of these particular drawings. He said he could not believe that Abbot had drawn them and contended that they had been drawn by an "inferior copyist." The "less said about it the better," Scudder wrote. "It was picked up at a book shop, bears the date 1830, and though Doubleday paid seven guineas for it, it is certainly not the work of Abbot, but of a very inferior copyist—some of the paintings being the merest daubs. It has scarcely the least value."[47] Scudder had been accustomed to seeing Abbot's earlier butterfly drawings and the elegant watercolor drawings in the "flora et entomologia" style (more on this style later); his rejection of these later drawings as "the merest daubs" is understandable. However, the "inferior

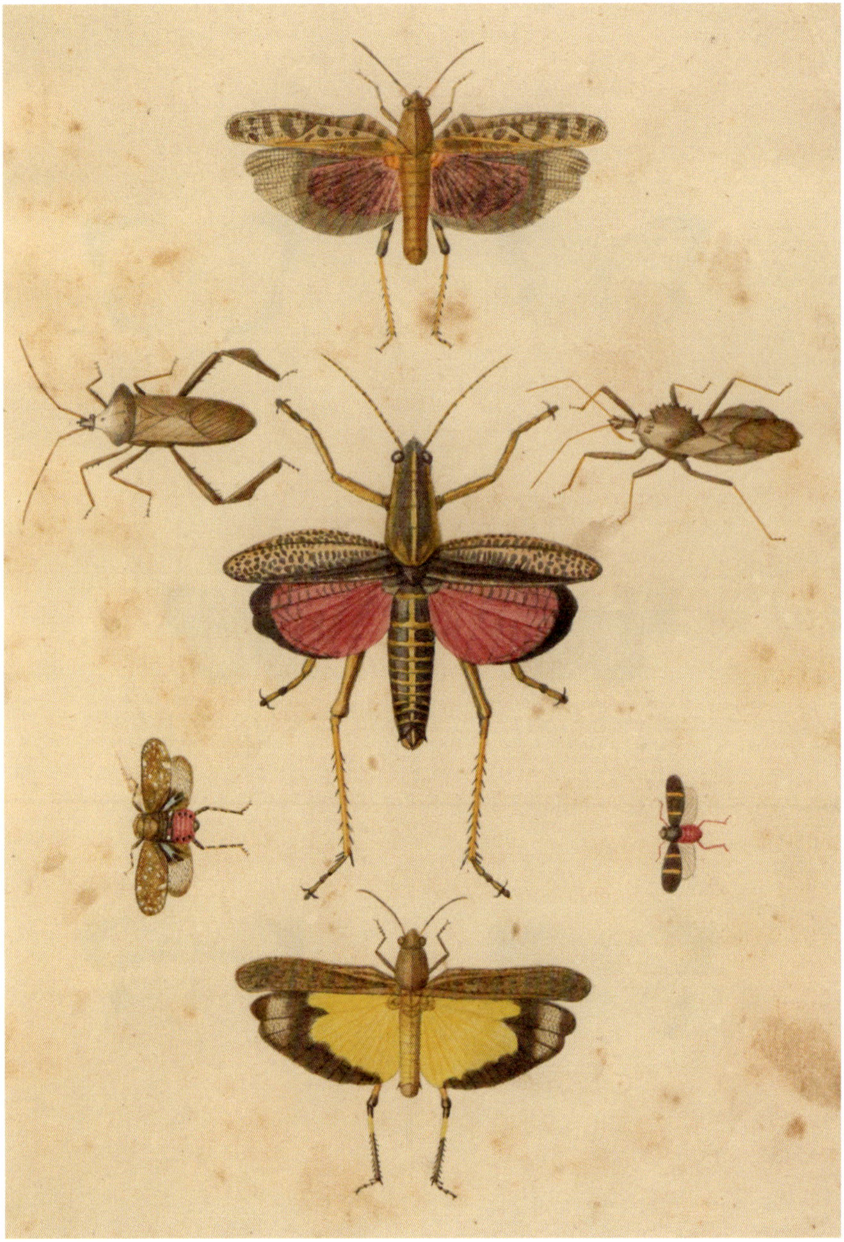

FIGURE 4.6.
Abbot, "Locusts" (*starting from top and moving left to right*), orange-winged grasshopper (*Pardalophora phoenicoptera*), giant leaf-footed bug (*Acanthocephala declivis*), North American wheel bug (*Arilus cristatus*), eastern lubber grasshopper (*Romalea microptera*) (*center*), sooty planthopper (*Poblicia fuliginosa*), two-lined spittlebug (*Prosapia bicincta*), yellow-winged grasshopper (*Arphia* sp.), det. J. K. Wadleigh, watercolor drawing, "Insects of Georgia" (1827), Stuart A. Rose Manuscript, Archives, and Rare Book Library, Emory University.

copyist" was indeed Abbot, who in his late seventies had reverted to the highly artificial style of representation that mirrored the display cases of cabinets in the early modern period.[48]

Although lacking the grace and elegance of his earlier drawings, these albums with their brightly colored gem-like beetles and fantastically colored crickets and moths are charming in their own way and within their own terms. Scudder's negative reaction to these drawings may have been intensified by the materials Abbot had used to make them. Instead of the fine British paper and pigments that Abbot had worked with throughout most of his career, these drawings of the late 1820s were painted on American-made paper, which was

FIGURE 4.7.
Abbot, penciled IDs were made most probably by T. W. Harris (*starting from top and moving left to right*), squash vine borer (*Eichlinia cucurbitae*), det. agreed upon and updated by J. K. Wadleigh, oak stump borer moth female (*Paranthrene asilipennis*), det. J. K. Wadleigh, oak stump borer moth male (*Paranthrene asilipennis*), det. J. K. Wadleigh, hummingbird clearwing moth (*Hemaris thysbe*), det. agreed upon and updated by J. K. Wadleigh, yellow-banded wasp moth (*Syntomeida ipomoeae*), det. agreed upon by J. K. Wadleigh, male and female eight-spotted forester moth (*Alypia octomaculata*), det. agreed upon by J. K. Wadleigh, watercolor drawing, "Insects of Georgia" (1830), MS Typ 426.5, Houghton Library, Harvard University.

inferior to British paper, with its high linen content, and the pigments in the paint are more metallic and opaque, closer to gouache, than the translucent paints of his earlier work.[49] What these drawings lack is that skillful handling of translucent paint that allows the white paper to shine through. This aquarelle technique, which Avinoff so admired, brought light into the drawing, creating the effect of depth, and contributed to the illusion of three dimensions. Abbot also seems to have added too much gum arabic to his watercolors, making some of these drawings from the late 1820s too shiny (see figure 4.3). Despite the diminished quality of his materials, Abbot was still able to draw insects accurately enough for entomologists to determine their genus and even species.[50]

Abbot's early drawings with their geometrical displays of insect bodies were, according to Avinoff, "masterpieces," while the geometrical displays of insects in Abbot's 1830 album were, according to Scudder, the product of "an inferior copyist." The disparity between Avinoff's and Scudder's estimations of the

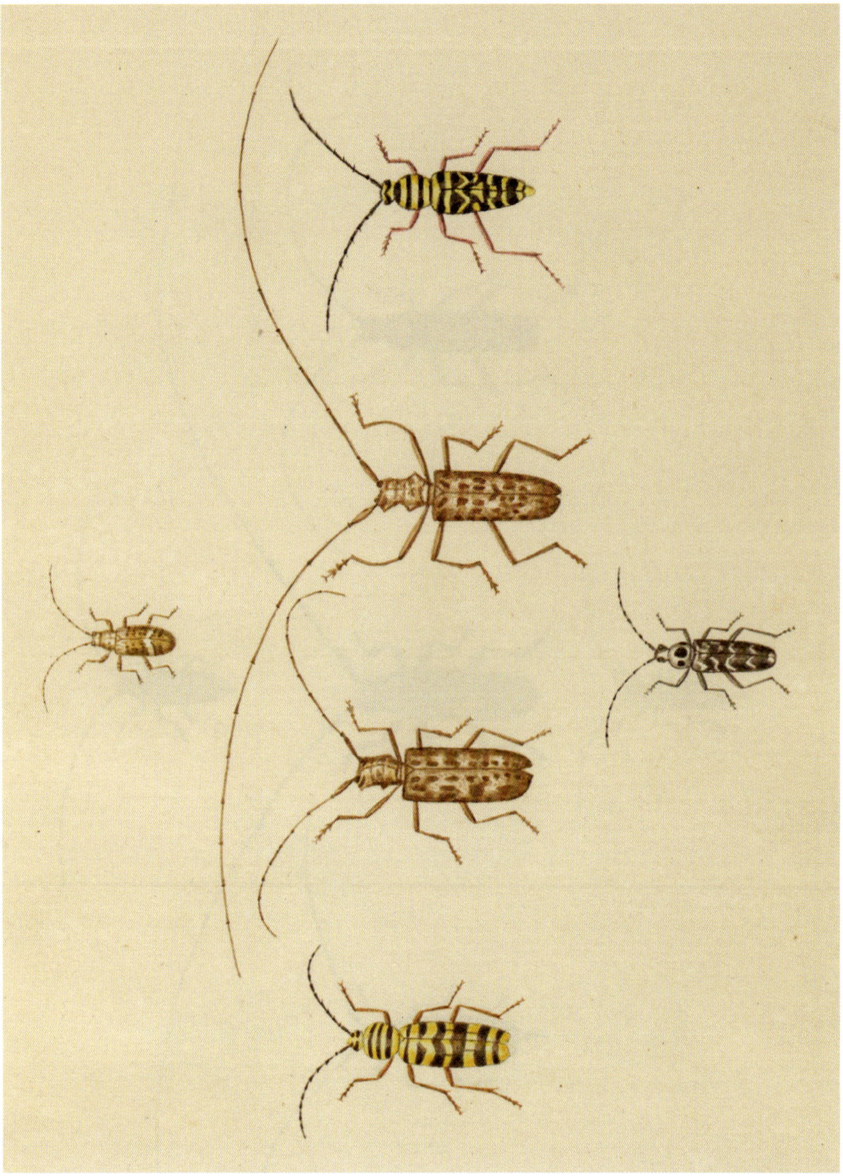

FIGURE 4.8.
Abbot, locust borer (*Megacyllene robiniae*), flat-faced longhorn (*Acanthocinini* sp.) (two middle, male and female), pine sawyer (*Monochamus* sp.), cottonwood borer (*Plectrodera scalator*), amorpha borer (*Megacyllene decora*), det. J. K. Wadleigh, watercolor drawing, "Insects of Georgia" (1830), MS Typ 426.5, Houghton Library, Harvard University.

value of these drawings speaks to Abbot's longevity and the rather poignant consequences of continuing to draw and paint into one's late seventies and early eighties. Scudder's disparaging comments may give some of us pause to contemplate whether we in our old age are inferior copies of our youthful selves.

Drury's Collection of Abbot's Drawings

Between the drawings of his youth and his old age that display insect bodies in tabular and geometrical patterns, Abbot produced drawings on quarto paper that ranged from depictions of a single figure to multiple figures on the page. Among these drawings are at least forty that I believe he made for Drury in the late 1780s and early 1790s. These drawings are special because they show Abbot experimenting with composition, moving away from decorative and tabular arrangements of insects to developing what became his dominant subject matter in quarto: the depiction of lepidopteran metamorphosis with larvae and host plants.

The whereabouts of Abbot's drawings owned by Drury had been a puzzle until about ten years ago, when I stumbled onto a cache of them in Harvard's Houghton Library. That Drury had such a collection of drawings is not surprising, since he encouraged Abbot to draw insects in his collection when the artist was a teenager. The silversmith also spent more than a decade coaxing Moses Harris into illustrating his insect collection. Clearly, Drury was invested in the visual representations of insects, especially those in his collection.[51] Drury's manuscript catalog of his own collection reveals that he possessed thousands of insect specimens that Abbot had collected in Georgia. We know Abbot sent Drury these specimens in boxes with hidden compartments in which the drawings and notes accompanying the specimens were stored.[52] The auction catalog for the sale of Drury's insect collection does not mention Abbot's drawings, although it lists fifty sets of uncolored prints from the three volumes of his *Exotic Insects* along with the copper plates, all of which his son was apparently eager to get rid of. Drury may have sold or given away Abbot's drawings as his health declined in the late 1790s, when he made plans to sell his collection. Drury put Edward Donovan, whose book on how to collect insects we encountered in the previous chapter, in charge of writing up the catalog for this sale.[53] Donovan wrote on the outer cover of the manuscript catalog that listed all of Drury's Lepidoptera: "These I received from Mr. Drury for whom I wrote the sale catalogue of Drury's Insects; these he gave me as I declined payment." "These" presumably refers to Drury's collection of butterflies and moths. During this process, Donovan may have acquired Abbot's drawings from Drury himself or in the aftermath of the silversmith's death. Proud of his relationship with Drury, Donovan eagerly purchased hundreds of Drury's auctioned specimens, boasting in his publications about the provenance of these specimens.[54] Whether Donovan or his heirs were responsible for depositing Drury's collection of Abbot's drawings in the British Museum we cannot know, but there they resided until 1869, when a curator sent them to a Harvard professor, who then bestowed them on the Boston Natural History Society.

Harvard University's Houghton Library possesses four albums of Abbot's insect drawings, with each album containing well over a hundred drawings. One of these volumes, MS Typ 426.2, contains a bookplate and prefatory material, though neither indicates that Drury owned it. What gives this album away as having been Drury's is the wobbly writing in black ink on the back of thirty-three of these drawings. Drury had transcribed verbatim from Abbot's notes the relevant information about the pictured insects. On the back of a drawing of three lovely beetles, Drury's transcription reads:

> Scarabeus Tityus, Male and Female. The Larva differs Little except in Size. Some feed on the insides of decayed trees and earth, rich with manure or dung. It breeds in Oaks, flies at Night & sometimes enters houses. It is scarce particularly the Male. taken about the 25th of May & as late as the 6th October. After death they turn black, but recover their Colour in a year or two & sooner when exposed to the Sun. They are called the Elephant beetle in Georgia. the Variety 2 is sometimes seen clouded.[55]

Abbot's original notes from which Drury had copied are lost, but because Abbot produced several iterations of the same drawings (more on this later), the notes he sent along with his drawings are remarkably similar across several sets of drawings. What Drury copied from Abbot's notes about these beetles, for instance, is repeated in notes that accompanied other sets of drawings, such as those in the Francillon Collection.

The distinctive scrawl on the back of these particular drawings tells us that they were in Drury's possession. I would not have recognized his handwriting had I not just returned from a trip to the Natural History Museum in London, where I had spent much of my summer reading Drury's correspondence and private papers. I had hopes that I would find mention of Abbot's name in two little memorandum notebooks, one dating from July 1794 to April 1796 and another from September 1801 to June 1805.[56] In the second notebook, I found notes Drury had made in reference to John William Lewin, a natural history artist whose sojourn in Australia Drury had sponsored. Drury was reminding himself to send a "caution to Mr. Lewin not to travel in the Interior of New South Wales also about sticking the insects." While I was disappointed not to find Abbot's name in these small notebooks, they turned out to be their own reward; they were completely charming, with Drury's recipes for homemade fruit wine, herbal elixirs for good health, and opium-laced medicine for what seems to have been a gallbladder complaint. My favorite entry is a list of items he packed for a two-week visit to a friend's country house: five nightcaps, three pairs of linen drawers, four shirts, two pairs of nankeen breeches and one pair of satin breeches, three pairs of silk stockings, four pairs of white cotton stockings, and assorted miscellaneous items such as a "shooing horn," two razors, a comb, and opium. This list was followed by a note to himself: "Take next time Slippers & Glister pipe. I need not have taken so many things." The second notebook dates from a time when he was frequently unwell and increasingly feeble, judging by his scratchy, up-and-down handwriting, which was quite different from his slanted, looping handwriting of the 1760s and 1770s. If I had not recently spent time reading Drury's recipes and lists of packed items, I might have not made the connection between Drury and the handwriting on the back of the drawings in the Houghton album. This unsteady handwriting can tell us when Drury transcribed these notes, probably toward the end of his life (he died in

FIGURE 4.9.
Abbot, "Elephant beetle" (male, male, female), eastern Hercules beetle (*Dynastes tityus*), det. J. K. Wadleigh, watercolor drawing, MS Typ 426.2, Houghton Library, Harvard University.

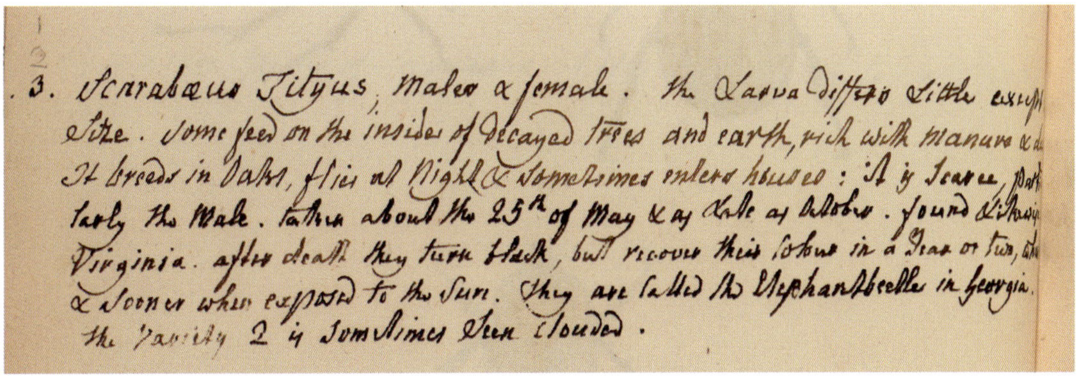

3. Scarabaeus Tityus, Males & female. the Larva differs little except
Size. Some feed on the inside of decayed trees and earth, rich with manure & d
It breeds in Oaks, flies at Night & sometimes enters houses: it is scarce, par
ticularly the Male. taken about the 25th of May & as late as October. found likewi
Virginia. after death they turn black, but recover their colour in a year or two, and
& sooner when exposed to the Sun. They are called the Elephantbeetle in Georgia.
the variety 2 is sometimes seen clouded.

FIGURE 4.10. Drury's annotation on the back of the Hercules beetle drawing, MS Typ 426.2, Houghton Library, Harvard University.

FIGURE 4.11.
Abbot, tooth-necked longhorn beetle (*Prionus pocularis*) (male and female), det. J. K. Wadleigh, watercolor drawing, Drury album, MS Typ 426.2, Houghton Library, Harvard University.

FIGURE 4.12.
Abbot, eastern tiger swallowtail (*Papilio glaucus*), det. agreed upon and updated by J. K. Wadleigh, watercolor drawing, Drury album, MS Typ 426.2, Houghton Library, Harvard University.

December 1803), when his handwriting got quite shaky, even though he could have had the drawings and notes in his possession from the mid-1780s, when he started receiving large shipments of insects from Abbot.

The drawings in this Houghton album that Drury owned depict Coleoptera, Vespa, and Lepidoptera. The drawings that have Drury's scrawl on their back are of beetles and wasps. His handwriting disappears on page 42 when the Lepidoptera take over the rest of the album. Fifteen of the thirty-five drawings of Coleoptera are presented as single insects on a sheet of quarto paper; thirteen drawings feature two different beetles, one above the other; two drawings have three beetles of the same species—male, female, and juvenile (see figure 4.9); and five drawings are beetle life histories with male, female, grub, and pupa lodged in a chunk of wood (see figure 4.11). After the beetle drawings are five drawings of wasps and hornets (see figure 2.9). These are followed by drawings of Lepidoptera figured in four ways: (1) as a single lepidopteran alone on a large sheet of paper (twenty-two drawings); (2) as two slightly different lepidopteran figures (some male/female?) on a single sheet of paper (twelve drawings); (3) as three lepidopteran figures arranged in a vertical line (one drawing); and

FIGURE 4.13.
Abbot, giant leopard moth (*Hypercompe scribonia*), det. J. K. Wadleigh, watercolor drawing, Drury album, MS Typ 426.2, Houghton Library, Harvard University.

(4) as part of an assemblage of lepidopterans, their larvae, and their host plants (ninety-seven drawings). (See figure 4.14.)

On the back of some of the Lepidoptera drawings, in particular those with host plants, someone has written identifications in pencil. These identifications were written not by Drury but possibly by Samuel Scudder, with a few references to Jean Baptiste Boisduval, about whom we will learn more in the next chapter.[57] Some damage seems to have occurred when the drawings were trimmed excessively and bound into an album; some of Drury's words are sliced off the outer edge, and others are tucked into the binding. This carelessness is something Drury would not have condoned, which leads me to attribute the binding to someone other than him. The binding was most likely done when the drawings were acquired by the Boston Society of Natural History in the 1870s.[58] With only a third of these drawings bearing Drury's annotations, it is tempting to suggest that this album is a late nineteenth-century compilation of various drawings by Abbot; however, the color, texture, and heft of the thick yellow paper are remarkably similar across the whole volume, suggesting that the drawings were all done at the same time by Abbot, who could have sent them to Drury in the late 1780s along with hundreds of insect specimens and accompanying notes, all packed up in those special boxes with their false bottoms to hide the drawings and notes. This particular volume could be one of the earliest sets of extant drawings that Abbot shipped to London from Georgia.

How did these particular drawings end up in one of Harvard's libraries? The bookplates and a letter inserted into the volume reveal quite a bit about how they made a return journey across the Atlantic. They were sent as a gift in 1869 to Professor Asa Gray of Harvard University from John Edward Gray (no relation), keeper of the Zoological Department at the British Museum. A few years later, in 1874, Professor Gray donated the album to the Boston Society of Natural History with the idea that these drawings would complement those by Abbot that this institution already held in its library. The Boston Society possessed at that time the aforementioned volume of "inferior" drawings that had belonged to Thaddeus W. Harris, and Asa Gray thought the society would appreciate having another album of Abbot's drawings. John Edward Gray had indicated that the British Museum did not need any more of Abbot's drawings, since the natural history division of the museum already possessed the stellar collection that had been amassed by John Francillon, the museum acquiring it in 1818 shortly after his death in 1816.

Giving away natural history objects, including artwork and specimens, was a common strategy that natural history museums deployed to offload unwanted items that had been donated or acquired and were no longer considered valuable enough to take up space, which was always in short supply. Some museums would even auction their unwanted collections that had been donated by naturalists.[59] Another form of exchange between museums and libraries was the swap. This is what occurred in 1915, when Harvard's Museum of Comparative Zoology gave the 1772 volume of Abbot's drawings to the Boston Society of Natural History in exchange for the society's copy of Jules Pierre Rambur's *Faune entomologique de l'Andalousie* (two volumes, 1837, 1840).[60] Gifting, swapping, and selling unwanted acquisitions were standard procedures among learned institutions, and as such, we need not read too much into these exchanges, except to note that neither the British Museum nor the Boston Society seemed to be aware that some of Abbot's drawings it had in its possession had been owned by Dru Drury, who, although dead for over seventy years at that time, still had a reputation as one of the founders of British entomology, a reputation that was enhanced by the updating and republication of his *Exotic Insects* by J. O. Westwood in the 1830s.[61]

Bound together with the drawings that bear Drury's distinctive scrawl are what may be Abbot's earliest concerted effort at picturing North American Lepidoptera with their host plants. He produced ninety-seven drawings in this format.[62] With this turn toward introducing host plants and larvae into his depiction of Lepidoptera, Abbot transformed his entomological portraits into something more complex: a flora et entomologia. In drawing a moth fluttering around its host plant and its caterpillar creeping up the plant's stem, Abbot chose to utilize a visual template forged a hundred years earlier, when Maria Sybilla Merian's intense study of lepidopteran metamorphosis led her to depict the butterfly or moth, its caterpillar, and the plant on which the caterpillar fed in the illustrations for her lovely book on European caterpillars, *Der Raupen wunderbare Verwandlung und sonderbare Blumennahrung* (Caterpillars, their wondrous transformation and peculiar nourishment from flowers) (Nuremberg, 1679, with the second volume appearing in 1683), and her famous masterpiece, *Metamorphosis insectorum Surinamensium* (The metamorphosis of the insects of Surinam) (Amsterdam, 1705).[63] Abbot borrowed Merian's incorporation of plant life into the drawing of insects for the illustrations in *Insects of Georgia*, a decision that delighted many of this book's reviewers. Adrian H. Haworth, author of *Lepidoptera Britannica* (1803), wrote in his enthusiastic review: "The Insects are charmingly represented in all their mutations [stages of metamorphosis]; and on such large and characteristic specimens of their food, the whole Plants as well as Insects being scientifically delineated and described, that this publication is to the full as valuable to the Botanist, as it is to the Entomologist: we never before beheld the sister sciences walk so closely, and so engagingly hand in hand, as in this interesting volume.—it is truly a Flora et Entomologia."[64] Haworth seems to have forgotten Merian's work in his enthusiasm for Abbot's skillful execution of not only lepidopterans but also the flowers, leaves, and snipped twigs that the larvae fed on. Also forgotten in Haworth's review were the early and mid-eighteenth-century publications by Eleazar Albin, Moses Harris (see figures 3.12, 3.13, and 3.14), and Benjamin Wilkes, whose illustrations pictured caterpillars in

Liquidambar styraciflua.

addition to moths and butterflies and sometimes with the plants they thought were hosts to the feeding larvae (figure 4.15).[65]

FIGURE 4.14.
Abbot, luna moth (*Actias luna*), watercolor drawing, Drury album, MS Typ 426.2, Houghton Library, Harvard University.

Abbot, fully aware of the work of these artists, self-consciously used their methods of presentation, but his drawings of butterflies and moths were more restrained than his forebearers' lively and exuberant renderings of insect life. While Abbot admired these other artists' drawings of Lepidoptera greatly, he also realized that the Baroque era embrace of the decorative over the precise rendition of these insects would not meet the needs of natural history collectors and naturalists, who, by the end of the eighteenth century, wanted insects depicted in accordance with Linnaean systematics. It is to these "drawings of the larva, pupa, and perfect insect of every lepidopterous species, as well as of the plant upon which it fed" that we now turn.[66]

The Flora et Entomologia Drawings

The words "Red haws," written in pencil, sit at the very top of a drawing of a flowering hawthorn around which flutters a tussock moth. This annotation was made by T. McDougal, the previous owner of the Hargrett albums of Abbot's drawings mentioned in the opening of this chapter. Nearly all of the drawings are similarly marked with the plants' common names. Noticeably interested in identifying the plants, McDougal transcribed their names along with those of the insects from Abbot's accompanying notes. His keen interest in the botanical part of these flora et entomologia drawings was shared by several of Abbot's contemporaries, including amateur and professional botanists. Abbot's friend Augustus Oemler, for instance, a Savannah apothecary who was also an amateur botanist, possessed what he thought was the finest set of Abbot's flora et entomologia drawings. Among these lovely drawings are many that are only partially painted. Apparently, Oemler had asked Abbot to "give the outlines of the plants in pencil only," lending some of these drawings a kind of ghostly quality, an absence within a presence.[67] The penciled outlines of the plants reveal Abbot's method of tracing in pencil an outline from a template onto paper, which is then painted with watercolors, functioning a little bit like a modern-day coloring book.

Oemler carefully annotated these drawings, writing on them with pencil the Linnaean names of the pictured plants.[68] Apothecaries in general were knowledgeable about plants and their medicinal properties, and Oemler must have thought Abbot's grasp of proper botanical nomenclature was lacking in comparison to his. Early on in their friendship, he had criticized some of Abbot's plant drawings for failing to depict the proper number of pistils and stamens, which was key to identifying them within the Linnaean system of classification.

FIGURE 4.15.
Benjamin Wilkes, "The Buff-Tip Moth" (*Phalera bucephala*), engraving, plate 43, *One Hundred and Twenty Copper-Plates of English Moths and Butterflies* (London, 1773), Nationalbibliothek, Vienna.

For this reason, Oemler may have preferred to identify and label the plants himself. (See figures 4.16 and 4.17.)

Abbot was modest about his knowledge of botanical matters, telling Swainson that he was "no Botanist, only an admirer of Nature[']s Beauties."[69] Clearly, these drawings pleased botanists and plant lovers alike. However, Abbot's reasons for including plants in his drawings were not simply to make pretty pictures of flowering plants that would appeal to both botanists and lepidopterists. His goal was to represent the life cycle of butterflies and moths, and plants were a crucial part of this process. The plants he included in these drawings were to the best of his knowledge those on which the various species of caterpillars fed.[70] His drawings of plants, however, depart ever so subtly from traditional botanical illustrations that present an idealized version of the plant. Such traditional illustrations place a snipped twig on a white background, freezing the plant's flower in a perpetual bloom, an image of timeless beauty. Abbot's introduction of insect life into the botanical illustration unsettles this stasis with an allusion to the movement of flying moths and creeping caterpillars.

While the lovely flowers and carefully rendered leaves lend much charm to these drawings of lepidopteran metamorphosis, what makes these mixed botanical and entomological drawings so intriguing is the way in which Abbot depicts the caterpillars. Often portrayed in the midst of eating leaves that already have chunks missing from them, Abbot's caterpillars are lively, even humorous, as they rear up and wave their upper bodies around, looking for the next leaf to munch on. Abbot depicts the caterpillars as active, while the butterflies and moths, even though drawn as if in flight, look almost too much like pinned specimens with their fully extended wings laid flat against the white background. Their spread and flattened wings were, of course, requisite for those who wanted to study the elaborate patterns on their wings as part of the process of identifying their species, but that very stasis reinforces the caterpillars' capacity for agency. These caterpillars navigate the space of the plate, creeping up the plant's stem or across a leaf, moving with purpose. Time passing outside the frame of the picture is evoked by Abbot's inclusion of the half-eaten leaves, which ask us to imagine what went on before the depicted scene. Movement across space and time lends these caterpillars a vitality that is reinforced by the mildly humorous allusion to the insatiability of these constantly growing and changing life forms. Abbot knew these caterpillars intimately and experienced firsthand their appetite for life.[71] (See figure 4.18.)

Abbot produced hundreds more of these flora et entomologia drawings. In 1813 Heinrich Escher Zollikofer began a correspondence with Abbot, inquiring about purchasing specimens and drawings, in particular those drawings that displayed the life history of Lepidoptera along with their host plants. Abbot responded in April to Escher's query by explaining that he was well acquainted with "the changes" of life and had drawn "about 350 species of Catterpil-

FIGURE 4.16.
Abbot, "Least Blue butterfly" (*Cupido comyntas*) (1805–25), watercolor drawing, Oemler's Collection, MS Typ 426.3, Houghton Library, Harvard University.

FIGURE 4.17.
Abbot, banded sphinx moth (*Eumorpha fasciatus*), det. J. K. Wadleigh (1805–25), watercolor drawing, Oemler's Collection, MS Typ 426.3, Houghton Library, Harvard University.

lars." Abbot offered to sell the Swiss collector "150 such Drawings as you desire Q[uarto] size in Colors," which he happened to have on hand. He explained that he had made these life history drawings for the Savannah Library but had not been paid yet, and he feared the deal was falling through. "The managers of the Savannah Library have offered me a Dollar a peice; Mr Oemler is Librarian, he is anxious the Library should purchase them, the bargain is not finally settled, I still have them in my own possession. I cou'd not therefore take less for these or any similar Drawings than the Library offers."[72] Not hearing back from Escher, who had returned to Zurich from New York, Abbot wrote him in December: "Not receiving an Answer to my last, has much discouraged me." He offered "to take less price for them" because they were "already done." Abbot worried that he had lost Escher as a customer, but soon after this initial confusion, they went on to establish a strong relationship that lasted more than two decades. Between 1816 and 1818, Abbot sent his Swiss correspondent 201 drawings, 174 of them featuring Lepidoptera and their host plants, and in the 1820s Abbot sent to Zurich yearly shipments of cork-lined boxes filled with specimens.[73] Escher had been especially interested in drawings that depicted the life histories not only of butterflies and moths but also of beetles, crickets, grasshoppers, and wasps, something that Abbot was well positioned to do, as he raised a diverse group of insects that included, surprisingly, the very destructive *Dermestes* beetles (See figure 4.21.).

In addition to the drawings owned by Drury, Oemler, and Escher, William Swainson purchased Abbot's drawings with the idea that he would produce a sequel to Smith and Abbot's *Insects of Georgia*.[74] Eager to take up what had been Smith's role in preparing *Insects of Georgia* for publication, Swainson had asked Abbot to send him "a series of drawings of all the species of Papilio and Sphinx which are not figured in Smiths work (these I should think would not be many) provided you can accompany them with drawings of their metamorphosis in the Caterpillar & Chrysalis state with the plants on which they feed, I would wish them if possible executed on good and strong drawing paper. . . . A few drawings of the other Genera of Insects would also be welcome if accompanied by their metamorphosis."[75] Abbot responded:

> I have commenced making a set of Quarto (large size) Drawings of the changes of Insects with notes, of such Insects that are not figured in Smiths Lepidoptera Insects of Georgia, indeed it is a continuation of that Work[.] I shall expect to be able to complete about 100 by the time I shall have your Collections of Insects ready to send You. I have always not have had less than 7^s 6d sterling a piece for such Drawings, but I am willing to take 6^s a piece for these. As I continue to make new discoveries, I can very readily make at least 200 such Drawings not figured in Smiths work, among them is many of the principal Insects both for size & beauty.[76]

FIGURE 4.18.
Abbot, "Orange White-Spot Moth, *Phalaena pellucide*," pink-striped oakworm moth (*Anisota virginiensis pellucida*), original drawing, TAB. LVI, Smith and Abbot, *Insects of Georgia* (1797). Courtesy of John Work Garrett Library, Sheridan Libraries, Johns Hopkins University.

Abbot sent him the drawings in 1818. Swainson was unhappy with them because he thought they were smaller than those Abbot had drawn for *Insects of Georgia*. (Swainson was wrong: The drawings were the same size, but the paper size was smaller than the book's printed pages.) Abbot agreed to work on a new set of drawings for him, but it all came to naught, as Swainson got caught up in other projects as he turned to the business of becoming a professional natural history author and illustrator.[77] Eventually, his career as a naturalist suffered a series of setbacks; the most serious of these was his failure to secure a position at the British Museum.[78] Without the financial support that such a position might have offered him, he was unable to sustain the intense work required to support himself and his family as a writer and illustrator, leading him to immigrate to New Zealand.[79] Despite Swainson's initial rejection of these drawings, he must have valued them greatly. He took them with him on his journey to the other side of the world, and he had begun the process of trying to identify the species Abbot had depicted, writing in pencil on a few of these drawings what he thought were the proper Linnaean names.

By my count there are extant well over a thousand of these lepidopteran life history drawings featuring caterpillars and their host plants if we include the original drawings for *Insects in Georgia*, now housed at Johns Hopkins University.[80] Because Abbot used templates to redraw his butterflies and moths, he was able to reproduce similar plates, and he sometimes recombined different plants with the same lepidopteran.[81] For example, Abbot's drawing in Escher's collection of a giant swallowtail butterfly depicts a snippet of an orange tree complete with an attached orange (see figure 4.19), while Abbot's drawing of the same butterfly and caterpillar in Swainson's and the Hargrett's albums shows a twig of a flowering Hercules-club, *Zanthoxylum clava* (see figure 4.20); both plants are in the cirtus family and host this swallowtail's larvae.[82] Abbot numbered his drawings and presumably his templates to correspond to specific species so that he could keep track of which drawings he had made for his clients. The Francillon orange tree swallowtail drawing carries the number 238, which would have corresponded to the numbered notes. Abbot would have used a ledger to keep track of what numbered drawing he had sent to a client. For example, he asked Stephen Elliot, a botanist living in Charleston, if he wanted more drawings: "Shou'd be glad to know if you want any more Drawings. . . . If you take any more please send me the Numbers of each in the Corner, that I may not mistake in sending what you have already got, and that I may send the Notes to each, to illustrate the history of each Insect."[83] Abbot made templates not only for Lepidoptera but also for host plants and for the beetles that he redrew and sometimes combined with plant life or displayed with their larvae (grubs) to depict metamorphosis. Life history drawings of beetles, crickets, and grasshoppers prefaced Swainson's, Escher's, and the Hargrett's drawings of butterflies and moths, as they did in the set of original drawings for *Insects of Georgia*.

Abbot produced these life histories over the course of three decades, from the early 1790s into the early 1820s. Swainson's drawings date from 1816 to 1818, and the Hargrett's are from a few years later; the drawings nearly duplicate each other.[84] Escher's collection of life drawings possibly dates a few years earlier, since Abbot and he had begun their correspondence in 1813.[85] Francillon's volumes 16 and 17 carry the date 1804 on the title page, but they actually may have been executed closer to 1810.[86] Harvard's ninety-seven flora et entomologia drawings bound with the beetle and wasp drawings that carry Drury's inscriptions date from well before 1803, the year of Drury's death. Those in the Drury album strike me as fresh and lively, much like the spectacular drawings Abbot sent to London in the early 1790s for publication. Some of the drawings made in the second decade of the nineteenth century look a little tired in comparison, but they still possess that charm derived from a mixture of pretty flowers and feisty caterpillars.

In tracing the provenance of Abbot's composite drawings, we have learned how they were used and why they were valued. Packed into a suitcase readied to flee a political conflagration or carried onboard a ship bound for New Zealand, Abbot's drawings elicited deep attachment in their owners. For Avinoff, Abbot's skills as an expert watercolorist combined with the observational talents of a practical naturalist made his drawings compelling for someone like Avinoff whose love of butterflies drove his scientific career and was expressed in his own artwork. For Swainson, Abbot's flora et entomologia drawings represented a moment in his life when he was positioning himself as an expert, taking up the task of editing a sequel to *Insects of Georgia* (1797), last completed by Sir James E. Smith, president of the Linnean Society.[87] This was a time when Swainson was on the brink of joining London's community of elite naturalists, an aspiration that was never fulfilled. Drury's scribbly transcriptions of Abbot's notes were his attempt to join the image of the insect with the living being, the notes providing the life history details he could never access in person, given his distance from the North American insects that fascinated him. The half-painted drawings Oemler owned offered him the opportunity to complete in his mind what was missing from the drawings: the empty outlines of the plants asking for his collaboration. Because these collectors found Abbot's composite drawings intellectually engaging and aesthetically pleasing, they were invested in protecting them against the forces of ruin and decay. In soliciting their owners' care, these charming drawings did their best to ensure their own preservation despite their material fragility as paper objects. More of these drawings are in private collections, and no doubt they too are treasured objects.[88] As lovely as these drawings are, they represent a sixth of Abbot's artistic output. The majority of his drawings are of individual insects on small pieces of paper, to which we now turn.

FIGURE 4.19.

Abbot, "Papilio Thoas," giant swallowtail (*Papilio cresphontes*), watercolor drawing, set a, p. 23, Escher-Zollikofer Collection, Zentralbibliothek Zürich.

FIGURE 4.20.

"Papilio Thoas," giant swallowtail (*Papilio cresphontes*), watercolor drawing, plate 7, "Moths of Georgia" (ca. 1817–20), Rare Book and Manuscript Library, University of Georgia Libraries.

FIGURE 4.21.

Abbot, "June Scarabaeus," common green June beetle (*Cotinus nitida*), det. J. K. Wadleigh, watercolor drawing, set a.2, Escher-Zollikofer Collection, Zentralbibliothek Zürich.

5

Francillon and LeConte

COLLECTING, CURATING, AND USING ABBOT'S SCIENTIFIC DRAWINGS

IN 1833 AUGUSTUS G. OEMLER, Abbot's old friend from Savannah, wrote to the artist, asking him whether he knew how many watercolor drawings he had done for John Francillon and John Eatton LeConte. Abbot replied that over the years he had worked for them, he had taken "no account of their [the drawings'] Number," and his answer could only be "merely guess work": "I think however I made about 350 Qr. [quarto] of the changes of the Insects for Mr Fn. [Francillon] & 2000 small ones, & 500 of Spiders, & must have made more small ones than that for Le Conte."[1] The "small ones" refer to drawings of a single insect on a small piece of paper, and "the changes of the Insects," a reference to metamorphosis, are the subject of the 350 quarto drawings that Abbot drew for Francillon featuring adult lepidopterans, larvae, pupae, and host plants, as discussed in the previous chapter.[2]

While Francillon and LeConte each collected over two thousand of Abbot's drawings, they had different motives and agendas for doing so. Francillon (1744–1816), the London silversmith, jeweler, and amateur naturalist, was the quintessential Enlightenment collector of natural history objects. He approached the insects he collected as would a connoisseur, mastering Linnaean nomenclature and the tools necessary to organize and identify what he collected. A member of the Linnean Society, Francillon was active in various natural history networks, displaying a generosity and affability that helped knit these early entomological communities together. Two generations younger, LeConte (1784–1860), a major in the U.S. Army, a topographical engineer, and a founding member of New York City's Lyceum of Natural History, used his collection of Abbot's drawings for scientific purposes.[3] He wrote and illustrated articles on beetles, turtles, frogs, and violets in the *Annals of the Lyceum of Natural History of New York*. For instance, his publication *A Monography of the North American Histeroides* contains illustrations based on his own drawings of beetles. As a distinguished naturalist and a gifted artist, LeConte became active in the Academy of Natural Sciences when he and his son, an entomologist, moved in 1852 from New York City to Philadelphia, where they thrived in this city's vibrant natural history culture.

In this chapter I describe Francillon's and LeConte's different approaches to acquiring and managing their collections of Abbot's entomological drawings, contrasting Francillon's careful curation of Abbot's drawings with LeConte's more scientific approach to handling them. Understanding how these two men engaged with these drawings will help explain what happened to them and where they are now. Attempting to account for the whereabouts of nearly five thousand drawings, I have tracked their movements from Georgia to Britain and France and back again to the United States, visiting university libraries, natural history museums, and scientific societies in search of where these thousands of drawings now reside. Recovering the paths along which Abbot's drawings traveled reveals the complexity of naturalists' networks and the tangled web of interactions, collaborations, and exchanges within entomological communities. We begin our comparison of Francillon's and LeConte's collecting practices with Francillon, who possessed nearly three thousand of Abbot's drawings, which, thanks to Francillon's efforts, are still extant.[4]

Francillon's Collecting Practices

In the 1790s, as Drury's health deteriorated and his interest in insect collecting waned, Francillon became Abbot's key client, accumulating over the next two decades thousands of specimens Abbot had collected and acquiring nearly three thousand of his drawings. In 1793 Francillon wrote to his friend John Leigh Philips (1761–1814), a wealthy Manchester silk manufacturer and natural history collector, to share the news of his recent acquisition:

> I have receiv'd from my friend at Savannah Georgia (the person who drew the hundred Birds) a wonderfull fine collection of Drawings of the Insects of that Country, they consist of all the Classes and Genera, and as extensive in each as possible he could collect the Insects to Draw from. The contents are 1021 sheets of Drawings, containing 1664 different Species, or 1883 figures[:] Viz. 70 Papilio, 780 Phalaena & 600 Coleoptera &c &c &c with a manuscript description of the natural History of each insect. They are all very Beautifull and correctly drawn.[5]

Impressed with the quality of Abbot's drawings and specimens, Francillon promoted Abbot's work among his entomological friends and associates, bolstering Abbot's sales within British naturalists' networks.

Acting as Abbot's agent, Francillon directed the flow of Abbot's specimens and drawings to particular people and organizations, becoming in the process somewhat of a gatekeeper, which may have caused some resentment. Francillon refused to sell Abbot's spider drawings to the esteemed arachnologist Baron Charles Athanase Walckenaer (1771–1852), as we shall see in the concluding chapter. He also refused to send Abbot's specimens to Swainson because Swainson wanted to be able to return the specimens if he did not like or need them. Francillon thought this was unreasonable and insulting to Abbot. He suggested that Swainson "send your order direct . . . [b]ut by no means in my Name as I do not wish to have any concern in it as I foresee it may create me much trouble, and Mr Abbot and myself have been many Years in great Friendship, having always found Him very correct in all his dealings, and no doubt he would be the same with you."[6] Soon after this exchange, Swainson did contact Abbot directly without Francillon acting as an intermediary.

In general, Francillon had Abbot's best interests in mind as he brokered sales of the artist's specimens and drawings. Francillon could be quite persuasive when trying to sell these items to his entomologically inclined friends and acquaintances. His letters to his friend Philips capture how persuasive the London silversmith could be. "I have just receiv'd a Box filled with fine Insects, and beautifully collected," Francillon announced to Philips and then proceeded to describe this "very fine collection and as fine as any I have ever receiv'd from my Friend, who still lives in Georgia near Savannah." Francillon urged Philips to act quickly, because other collectors were quite interested in this particular box containing "between 7 and 800 Insects": "I think they are worth your attention, no person has seen them but myself, and I shall be much oblig'd to you for your Answer as soon as possible as two Friends in London would take them immediately were either of them to see the collection. But I wish you to have the preference."[7] When Philips expressed his ambivalence about expanding his collection to include non-European insects, Francillon began his second pitch with flattery, warning Philips that "collecting a few English specimens in your own garden will never be sufficient to entertain a mind like yours":

> The collection is so fine I think it a great pity you should not have it and as you say in the conclusion of your Letter, you will like to have them, if it contains a Variety of Genera, and not made up of small moths, and that you will leave it up to me to determine. Therefore I have thought proper to send it by tomorrow Wednesdays Waggon. . . . I hope you receive it safe, and meets with your approbation. Pray send your servant for it at the time the Waggon arrives to prevent the Box being ruffly handled.[8]

Francillon succeeded in placing Abbot's specimens and drawings with Philips, as well as convincing him to purchase a bespoke mahogany cabinet, complete with thirty-six drawers lined with cork, to house these precious items.[9]

A decade later, Francillon also persuaded Philips to inquire if the Manchester Library would like to purchase Abbot's insect and bird drawings: "I have sent you the 90 Drawings of Insects &c &c by Mr John Abbot of Savannah at 8/6 each for your and the Gentlemen of the Manchester Library's Inspection upon sale or return and hope they will be approved of."[10] Some library members were not entirely happy with Abbot's use of fanciful vegetation, stumps, and moss as props for the birds to perch on, which was regarded as an old-fashioned approach to ornithological drawings. However, they chose to keep the bird drawings and even considered buying more if Abbot would drop the stylized plants. Surprisingly, the gentlemen of the Manchester Library turned down the offer of Abbot's drawings of the lepidopteran life cycle because they thought too many of these images were repeated in Smith and Abbot's book, *Insects of Georgia*, which the library already possessed. Francillon was disappointed, advising "the Library to sell the work [the book] they now have, and purchase original Drawings that will be so much more extensive." Francillon had recently acquired 245 drawings of lepidopterans with larvae and host plants, which he had bound up in two volumes (most likely volumes 16 and 17 now in the Natural History Museum). He thought the library was missing an opportunity to own these original and rare drawings, "a very fine work almost unique, as there will be only mine and theirs so extensive in being."[11] He did, however, acknowledge that "the Gentlemen of the library have the right to act as they please."[12] The Francillon-Philips correspondence held in the British Library does not include a discussion of an important transaction that Francillon may or may not have negotiated: Manchester Library purchased well over five hundred of Abbot's small, single-figure spider drawings, as attested by the bookplate affixed to the inside boards of this amazing volume, now held by the Hargrett Library.[13]

Francillon did not profit financially from the sale of Abbot's specimens and drawings. His motives for promoting Abbot's work were to bring pleasure to his friends and to promote their commitment to the study of natural history.[14] His zeal for natural history collecting was boundless; he supported those whose enthusiasm seemed to be flagging and nurtured new converts, showing them London's best insect collections and introducing them to prominent members of London's naturalist community.[15] He engaged in these activities because he was a consummate collector of natural history objects, and a huge part of collecting was managing the flow of these objects within naturalists' networks. In the 1780s, when he and Philips were sending each other gifts of lepidopterans they had caught and raised, Francillon asked Philips if he would do him a favor: "I must beg of you to procure me some more of the *Hero Papilio* as I want to serve a few friends in London who are very desirous to have them and have not in their power to travel so far for them."[16] In exchange for butterflies (and game) from the North, Francillon reciprocated with small gifts: pins for

FIGURE 5.1.
Abbot, "Sphinx," huckleberry sphinx (*Paonias astylus*) and twin-spotted sphinx (*Smerinthus jamaicensis*), det. J. K. Wadleigh, vol. 7, drawings 10 and 11, Francillon Collection, Archives and Library, Trustees of the Natural History Museum, London.

setting butterflies, cork for lining drawers, seeds from Australia, natural history auction catalogs, and entomological books from Europe. He readily offered help identifying specimens, telling Philips to send him the specimens that were giving him trouble: "Don't be afraid to impose on me, that cannot be, for I am always ready with pleasure to assist friends."[17] When Francillon sent Abbot's specimens and drawings to Philips in the 1790s and early 1800s, he was building upon a decades-old infrastructure of exchange while also elevating the quality of the objects that circulated within this collecting culture.

Francillon's triumph as a natural history collector has to be his collection of Abbot's drawings. Its sheer size is remarkable, but more importantly, he practiced techniques of curation that enabled the drawings to survive intact into the present. Excited about the prospect of organizing these drawings, Francillon told Philips about his plans: "I intend to bind them up in an elegant manner with blanck leaves of fine writing paper between each drawing to write the names and description opposite to each insect. I think I cannot bind them into less than ten Volumes, Quarto without making each book too thick."[18] Bound in lovely red morocco leather with gold volume numbers embossed on their spines, seventeen volumes are filled with Abbot's drawings of moths, cicadas, wasps, flies, and beetles, to list only a few of the thousands of arthropods he drew, the majority depicting insects as if they were set specimens awaiting examination by attentive collectors and naturalists. A typical page in one of these volumes (with the exception of volumes 14, 16, and 17) has one to three small drawings, each portraying one life-sized insect, pasted onto a blank sheet of folio-sized wove paper. Beneath these drawings are pasted slips of paper on which are transcribed Abbot's notes about the pictured insects (see figure 5.1). Volume 14 is filled with geometrically arranged spider drawings with notes written on the back of the preceding drawings; likewise, volumes 16 and 17, which feature life histories, have notes similarly placed. These neatly presented pages represent countless hours of work.

Using Whatman paper, Francillon transcribed hundreds of pages of Abbot's notes, which provided information about the pictured insects as living creatures. He edited and revised the original notes, in some cases turning phrases into sentences, rearranging the order of the sentences, and making small changes in vocabulary. On the whole, his transcriptions retain information about North American insects as living creatures, which, as we learned in chapter 2, was scarce in European centers of natural history inquiry. Even more daunting than transcribing thousands of Abbot's descriptions was the intellectual work of arranging the drawings, which entailed deciding how to organize the images according to the most recent Linnaean systems of classification. Francillon explained to Philips how he went about the process of identifying the arthropods that Abbot had drawn: "I have selected all the names I can by going thr' many

10

+++50.

11

△ 14.

N°50. Sphinx. Taken in Summer, see Quarto Drawings N° 116. called in Georgia the Cherry Laurel Sphinx.

N° 14 △ Taken in April, it is very rare, Sphinx

authors Viz. Lin, Fab, Cramer, Drury, Olivier, &c &c, and find that not one tenth part of them are described."[19] For the 90 percent of Abbot's drawings that were of insects not yet described by taxonomic authorities, he arranged the drawings by genus or family, putting like with like. He placed, for example, the drawings of different species of ichneumon wasps together for comparison and for future study and classification by more expert taxonomists, such as the entomologist Johann Christian Fabricius, Linnaeus's star student, who visited Francillon's collection in 1787: "I have had the pleasure of seeing Mr Fabricius in London. He was with me one day last week, ten o'clock in the morning till nine at night and described many of the Lepidoptera."[20]

Pictured among these thousands of drawings are colorful swallowtail butterflies and hawkmoths, elegant cicadas and dragonflies, and impressive scarab beetles and orb spiders. These beautiful arthropods are matched by drawings of creatures commonly considered more menacing, such as wasps, ticks, mosquitos, flies, centipedes, and black widow spiders. These striking images are a small fraction of the drawings Francillon had collected, which included hundreds of drawings of small unprepossessing Coleoptera and page after page of small brown moths that only a lepidopterist could love. Volume 14's spider drawings and volumes 16's and 17's display of mostly lepidopterans with their larvae and host plants are on quarto-sized paper, while approximately twenty-three hundred insect drawings are on small slips of paper.

According to W. F. Kirby (1844–1909), an assistant at the British Museum from 1879 to 1909, "the contents" of Francillon's volumes "are as follows":

Vols. 1–4. Coleoptera [803 watercolor drawings].
5. Orthoptera, Hemiptera, Homoptera and Heteroptera [289 watercolor drawings].
6. Lepidoptera, Rhopalocera [100 watercolor drawings].
7–11. Lepidoptera, Heterocera [616 watercolor drawings].
12. Neuroptera, Hymenoptera [277 watercolor drawings].
13. Diptera [273 watercolor drawings].
14. Arachnida [117 watercolor drawings].
15. Myriopoda, Mallophaga, Acarina, Crustacea, and Lepidoptera (transformations), &c. [20 watercolor drawings].
16. Portrait [Abbot's self-portrait], Orthoptera, Coleoptera (transformations), and Lepidoptera (transformations) [130 watercolor drawings].
17. Lepidoptera (transformations) [133 watercolor drawings].[21]

The depth, range, and artful organization of Francillon's collection of Abbot's drawings constitutes a curatorial triumph.

FIGURE 5.2.
Abbot, "Libellula," (male, female) seepage dancer (*Argia bipunctulata*), det. J. K. Wadleigh, vol. 12, drawing 45, Francillon Collection, Archives and Library, Trustees of the Natural History Museum, London.

137

+++ 614.

138

+++ 618.

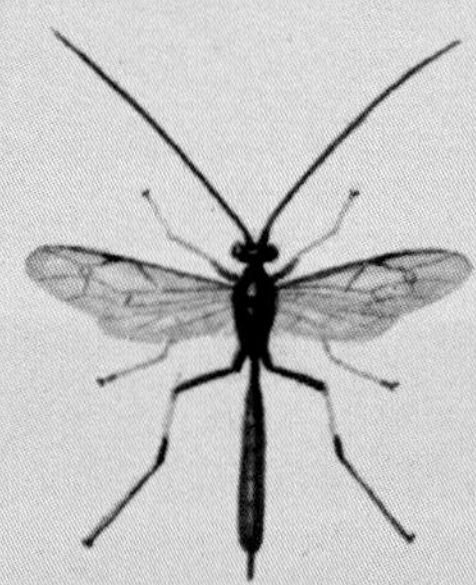

139

+++ 617.

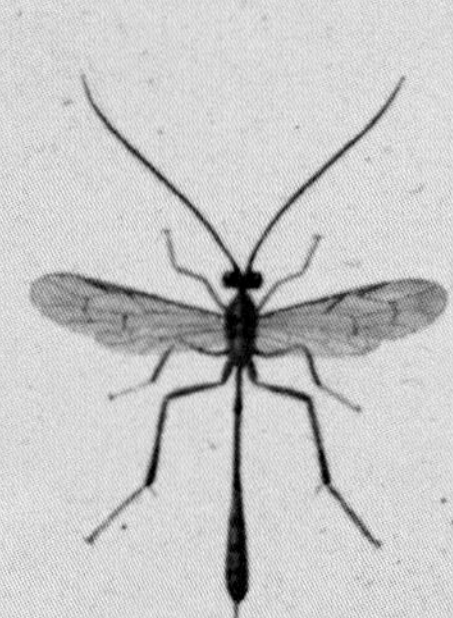

To understand what Francillon's curation of Abbot's drawings entailed, it is helpful to draw on book historian Elizabeth Yale's study of early modern scientific manuscripts and her important distinction between amassing papers and collecting manuscripts. Collectors, she argues, transformed piles of papers into manuscripts and in doing so created usable archives through "cleaning them up and reordering them, transcribing sections that were difficult to read, and binding them into books arranged by genre or topic."[22] The labor required to turn papers into an archive is extensive, often lasting decades and involving the employment of several hands. The care with which Francillon collected, organized, and displayed these thousands of drawings suggests that he valued them as precious objects that needed to be preserved. He protected their future as a collection by making arrangements for the seventeen volumes to be deposited in the British Museum upon his death rather than joining his insect specimens on the auction block, where his volumes could have been broken up and the drawings sold off piecemeal, suffering the vagaries of the marketplace.[23] Francillon's curatorial practices, unlike LeConte's, ensured, as much as possible, that his collection of Abbot's drawings would remain intact, weathering threats of dissolution and disappearance.

FIGURE 5.3.
Abbot, "Ichneumon," Ichneumonid wasps (Ichneumonidae spp.), det. J. K. Wadleigh, vol. 12, drawings 137–39, Francillon Collection, Archives and Library, Trustees of the Natural History Museum, London.

"Some Publications in View"

By the mid-1810s John Eatton LeConte had become Abbot's main client in the United States. In December 1816 Abbot wrote to Swainson, explaining that with Francillon's death he had only one client now, LeConte (whose last name is spelled variously, and not just by Abbot): "I am under no particular engagement to any person to collect or Draw for, but only to M^r Le Compte a Gentleman who has Plantations in Georgia, but is now at the Northward—he has a Taste in General for Natural history but seems mostly attached to Botany—for him I have Drawed and am continuing to Draw a large Collection of Insects. I believe he has some publications in view."[24] LeConte, "a man of science," probably met Abbot in Savannah, which was not far from where LeConte's brother, Louis, resided on Woodmanston, a rice plantation he had inherited from their father.[25]

Abbot estimated in the mid-1830s that LeConte had purchased two to three thousand drawings from him over the course of two decades. In 1835 Abbot explained to Thaddeus W. Harris that the majority of these drawings were single specimens drawn on a small piece of paper so that LeConte could "class them as he received them," meaning the small pieces of paper could be easily rearranged like filing cards as he tried to arrange them according to Linnaean systematics.[26] Abbot supplied these details about his financial arrangements with LeConte: "My charge to Mr. Le Conte for my Drawings & for whom I continue to draw

FIGURE 5.4. Abbot, "Nymphalis disipus," viceroy butterfly (*Limenitis archippus*), engraving, plate 55, Boisduval and Leconte, *Histoire générale et iconographie des lépidoptères et des chenilles de l'Amérique septentrionale* (1833), Robert S. Cox Special Collections and University Archives Research Center, UMass Amherst Libraries.

for every Year, the size of which I have figured at the end of this Letter is .16 for a dollar, they mostly are small Insects, & many are Minutia, & if small & will shew better, something magnified, he must be in possession of 2 or 3000, of them if the size of the paper & the Insect requires it, a larger price in proportion, 25 cents, but never exceeding 50 cents."[27] In 1834, when Harris suggested to Augustus Oemler, Abbot's friend, that Major LeConte supported Abbot by employing him to draw for him, Oemler bristled with indignation: "As for to be *supported* by Major Le Conte that is not so, except the petty allowance of Six and a quarter Cents for the drawing of an Insect (be it a flea, or a Bombyx

Polyphemus) with a full account of its natural history, may be called such."[28] Despite LeConte's standing order for drawings, Abbot was unable to support himself on what LeConte paid him. Oemler complained again to Harris in 1851 about LeConte's treatment of Abbot: "Le Conte brags a great deal that he 'maintained Abbot for several years.' How did he do it? He allowed him for each insect, large or small, in nature and a correct portrait of it, the enormous sum of Six and a quarter Cents, and after all, disputing the last bill of the old man, thus cheating him."[29] Oemler sensed in LeConte a lack of respect for Abbot and his abilities, LeConte regarding the artwork as raw data that he, as a man of science, could transform into scientifically valuable knowledge. As a result, an air of disregard haunts LeConte's attitude toward Abbot, a disregard that had an impact on how he treated Abbot's artwork, as we shall see.

To trace the whereabouts of LeConte's collection of Abbot's drawings, let's start with those that LeConte brought to Paris in 1828. Abbot was right that LeConte had "some publication in mind," referring to the drawings of Lepidoptera that he had made for LeConte in the 1810s and 1820s. The publication that LeConte was gathering drawings for was *Histoire générale et iconographie des lépidoptères et des chenilles de l'Amérique septentrionale* (1829–33), coauthored by Jean Alphonse Boisduval (1799–1879), a physician, a highly regarded lepidopterist, and a prolific author of several books describing Lepidoptera from all over the world. LeConte's main contribution to this book was Abbot's drawings and notes, which Boisduval used as source materials.

Most copies of Boisduval and LeConte's book are found today in the United States in natural history museums, scientific organizations, and university libraries, some of which were associated with agricultural programs. For instance, one of the copies I examined has a bookplate announcing that it had been the property of Massachusetts Agricultural College, founded in 1863 and a former iteration of what is now the University of Massachusetts, a land grant university.[30] The number of copies of this book held in land grant universities suggests that this book, unlike *Insects of Georgia*, was not a luxury item designed for wealthy natural history enthusiasts and lepidopterists of means; its readers extended beyond European collectors of North American Lepidoptera to people who needed practical information about the butterflies and moths that could possibly impact their agricultural pursuits. Unlike *Insects of Georgia*'s elegant illustrations printed in folio on fine Whatman paper, Boisduval and LeConte's book is smaller, published in octavo and in what are called "livraisons," which are bound portions of the book that could be rebound together once all were published. Printed on sturdy pages, this book, made up of twenty-six livraisons, was designed to be used rather than displayed.

Sixty-two illustrations in *Histoire générale et iconographie* identify Abbot as the illustrator, bearing his name at the bottom of the plate.[31] This book's illus-

FIGURE 5.5. Abbot, "Black veined orange butterfly," viceroy butterfly (*Limenitis archippus*), watercolor drawing, Escher-Zollikofer Collection, Zentralbibliothek Zürich.

FIGURE 5.6. Abbot, "Papilio Troilus (femelle)," spicebush swallowtail (*Papilio troilus*), engraving, plate 10, Boisduval and Leconte, *Histoire générale et iconographie des lépidoptères et des chenilles de l'Amérique septentrionale* (1833), Robert S. Cox Special Collections and University Archives Research Center, UMass Amherst Libraries.

trations are laid out in a mix of formats, with fifteen in the flora et entomologia style displaying the lepidopteran life cycle with imago, caterpillar, chrysalis, and host plant. In etching these plates, Boisduval's engravers rearranged elements from Abbot's compositions, compressing space and squeezing the figures onto pages half the size of *Insects of Georgia*'s, resulting in a loss of grace and elegance.[32] Another forty-five of the illustrations credited to Abbot feature one or two adult butterflies with larvae and pupae, an unusual composition for him but one that LeConte had specifically requested.[33] Three Abbot-based illustrations feature a single butterfly. Of the remaining fifteen illustrations, one is credited to LeConte, three to the engraver Paul C. R. C. Duménil, and eleven to Émile Blanchard, a gifted French illustrator who worked closely with Boisduval.[34]

French readers criticized Abbot's drawings for not depicting lepidopteran abdomens and legs properly, features that French lepidopterists regarded as important. In the preface to the tenth livraison, Boisduval apologized for these errors and explained that Abbot's focus had been on wing patterns, which is key to identification, and for this reason he had depicted the butterflies with their wings fully extended and not folded in their more natural position when at rest, which French readers apparently would have preferred. Some of Abbot's figures were reworked by Boisduval's artists in response to this critique. Despite these criticisms, Boisduval praised Abbot's contributions to the study of North American Lepidoptera as an artist and as a careful observer of their behavior and their appearance.

When LeConte returned to the United States, he left Abbot's drawings behind in Paris with Boisduval, who considered them precious enough to keep for further use.[35] Boisduval and Achille Guenée (1809–80), a fellow entomologist, mined Abbot's drawings of caterpillars, for instance, using them for the images in the atlas of their multivolume *Histoire naturelle des insectes; spécies général des lépidoptères* (1836–58).[36] Boisduval also shared his Abbot materials with the American lepidopterist Samuel H. Scudder (1837–1911), who visited him in Paris sometime in the late 1860s or early 1870s. Boisduval allowed Scudder to copy the larvae and pupae pictured in several of Abbot's drawings.[37] He also gave Scudder the notes that had accompanied three sets of Abbot's drawings, Scudder writing proudly on the first page: "Given me by Dr. Boisduval." Scudder was pleased to be the recipient of these notes: "Dr. Boisduval was good enough to present me with three series of manuscript notes entitled 'Notes to the Drawings of Insects,' all written in Abbot's own hand, and comprising twenty-seven foolscap pages, rather closely written, and describing the changes of two hundred and one species; of these thirty-eight are butterflies."[38] Boisduval's gift of natural history manuscripts to a fellow lepidopterist who would appreciate them is typical of the way in which drawings and other natural history manuscripts and objects circulated within naturalists' networks of exchange and knowledge production.

Where are the drawings that accompanied Abbot's notes that were gifted to Scudder by Boisduval? For years, scholars have assumed that LeConte provided Boisduval with the drawings featuring moths and butterflies with their caterpillars and host plants in the genre I have been referring to as flora et entomologia. The assumption has also been that these drawings were the basis of this book's illustrations and that they had disappeared. Did Boisduval keep them and then give them away, as he had the notes? The key to unraveling this mystery over the whereabouts of these drawings lies in Switzerland, where John Calhoun, a lepidopterist and expert on the butterflies in Abbot's drawings, has recently discovered the "lost" drawings that feature metamorphosis and host

FIGURE 5.7. Abbot's drawings were the basis for all but one of the caterpillars on plate 2 in the *Atlas*, J. A. Boisduval and A. Guenée, *Histoire naturelle des insectes; spécies général des lépidoptères* (1858), Hargrett Rare Book and Manuscript Library, University of Georgia Libraries.

plants. According to Calhoun, the drawings depicting the lepidopteran life cycle had been borrowed by Boisduval from Abbot's Swiss client, Heinrich Escher, and these drawings, not LeConte's, served as the basis for the illustrations with host plants.[39] Although these drawings were eventually returned to Escher and are now in the Zentralbibliothek Zürich, it seems that the notes that had originally belonged to Escher were not returned to him and were instead gifted to Scudder.

If Abbot's illustrations with host plants had belonged to Escher and were not among the illustrations LeConte had brought to France, what did LeConte bring to Paris? Abbot created the drawings brought to Paris according to LeConte's specifications: on small pieces of paper, without host plants, and depicting the caterpillar, pupa, and adult. These remained in Paris for a good while after LeConte's departure. Again, we are indebted to Calhoun's detective work, for according to him, in 1876 "Boisduval presented his library, ostensibly including these drawings to good friend and fellow Parisian lepidopterist Louis M. A. Depuiset. Depuiset organized all of Boisduval's assorted illustrations sometime before his death in 1886."[40] These drawings then found their way to the marketplace in the 1920s, when they were bought and disappeared into a private collection. They resurfaced in 1963 at a Sotheby's auction, where they were bought by a rare book dealer who sold them to the University of South Carolina's Thomas Cooper Library.[41] These drawings, used for forty of the illustrations in Boisduval and LeConte's book, feature two adults that are often presented in two poses, one with wings spread as if pinned and the other at rest in profile. The caterpillar hovers over the two adult figures, sometimes resting on a twig and sometimes presented with its pupa. This particular arrangement of figures was drawn on paper cut into rectangles of ten by six inches. True to LeConte's request, the layout of these drawings is unlike most of Abbot's work. (See figures 5.6 and 5.8.)

Another mystery involving Abbot's drawings is alluded to by Scudder's scrawl in black ink at the top of the list of butterfly drawings. Inserted into his line "Abbot's Drawings in Boisduval's possession" is the phrase in pencil "or LeConte's," which is placed right after Abbot's name. The line then reads: "Abbot's or LeConte's Drawings in Boisduval's possession." This annotation raises questions about who painted what. LeConte, a naturalist with a taste for botany as well as entomology and herpetology, was also a talented natural history artist. He drew elegant watercolors of turtles, frogs, beetles, and wild violets. Scudder's insertion "or LeConte's" opens up the possibility that LeConte may have contributed or intended to contribute several of his own drawings to the Boisduval volume. Only one of the seventy-eight published illustrations for *Histoire générale et iconographie* credits LeConte as its illustrator, but this does not preclude the possibility that LeConte intended to contribute more of his

own artwork if the project had extended beyond the livraisons that made it into publication. Scudder's insertion of LeConte's name suggests that LeConte may have been one of the artists along with Abbot responsible for the drawings in Boisduval's possession. This suggestion raises important questions about Abbot and LeConte's relationship and the potential for confusion over attribution of the work Abbot did for him.

FIGURE 5.8.
Abbot, monarch butterfly (*Danaus plexippus*), watercolor drawing, Irvin Department of Rare Books and Special Collections, University of South Carolina Libraries.

The University of South Carolina purchased what had been described by the dealer to be Abbot's drawings for the Smith collaboration, *Insects of Georgia* (1797), and then later were thought to be Abbot's drawings for the Boisduval and Leconte volume. However, recently doubt has crept into whether all of the 149 drawings can be attributed to Abbot. LeConte may have been responsible for at least twenty-eight drawings, some featuring individual caterpillars and others combining moths with their larvae.[42] The argument for LeConte as the artist of some of these drawings comes down to questions of style, specifically, the handling of paint. Abbot's aquarelle technique, according to Andrey Avinoff, is key to understanding what makes his watercolor drawings so charming and effective. By using diluted washes and holding back the application of paint, Abbot brought light into his watercolors while also creating depth in the image.

FIGURE 5.9. Abbot, hummingbird clearwing (*Hemaris thysbe*), watercolor drawing, Irvin Department of Rare Books and Special Collections, University of South Carolina Libraries.

He also used tiny dots of paint to mimic the feathered surface of lepidopteran wings. In the moth drawings that are now suspected of being by LeConte, the paint has been laid on heavily in strokes, layer over layer, a technique associated with the application of gouache, which has the effect of flattening the image. This is not to say that these drawings are inferior in style; they are just different from Abbot's, with their emphasis on flat patterned surfaces (see figures 5.9 and 5.10). Another twenty-three drawings have been attributed to various artists in Boisduval's employ, leaving the grand total of ninety-nine drawings that can be comfortably attributed to Abbot. Forty of the fifty-five butterfly drawings in this collection were published in *Histoire générale et iconographie*, while none of the moth drawings by Abbot (or LeConte) made it into print, as Boisduval and LeConte's plans to publish a continuation of the *Histoire générale et iconographie* came to naught.[43]

This concern over attribution applies also to another set of drawings that LeConte brought to Paris, drawings I believe are among those small ones that Abbot drew for LeConte. The University of Connecticut's rare book library possesses two bound albums of 450 small drawings of moths. On the leather spine of each volume is the title embossed in gold: "John Le Conte Lepidoptera of North America." The inside boards are beautifully papered in marbled

FIGURE 5.10. John Eatton LeConte, hummingbird clearwing (*Hemaris thysbe*), watercolor drawing, Irvin Department of Rare Books and Special Collections, University of South Carolina Libraries.

FIGURE 5.11. Abbot, polytemus moth (*Antheraea polytemus*), watercolor drawing, Irvin Department of Rare Books and Special Collections, University of South Carolina Libraries.

Macrogl. Gaurae Smith

purple and yellow, to which is affixed a decorative bookplate: "Exlibris Doctoris Boisduval." Pasted on the first page is a small piece of paper with a penciled note in French declaring that these original drawings were by John LeConte for his work on North American Lepidoptera in collaboration with Dr. Boisduval.[44] That Boisduval wrote this note is unlikely, given that it is in the third person. It was probably written by Depuiset as part of his attempt to bring order to Boisduval's library, which had been bequeathed to him. That could account for the error in assuming that LeConte was the artist who drew these little moths. Of the 450, 23 are signed J.L.C. (John [Eatton] LeConte), which suggests that these may be the only drawings by LeConte in these two volumes.[45] The rest of the drawings certainly seem to me to be Abbot's. Some of them are recognizable, having appeared in other formats and collections.[46] We can assume that most of the 450 small drawings of moths that LeConte left with Boisduval in Paris and that are now held at the University of Connecticut are among those two to three thousand drawings Abbot made for LeConte.[47]

FIGURE 5.12. Abbot, proud sphinx (*Proserpinus gaurae*) and Delaware skipper (*Anatrytone logan*), det. J. K. Wadleigh, drawings 17 and 18, in "John Le Conte Lepidoptera of North America," Archives and Special Collections, University of Connecticut Library.

FIGURE 5.13. Abbot, oak gall borer moth (*Synanthedon decipiens*) and (female) tupelo clearwing moth (*Synanthedon rubrofascia*), det. J. K. Wadleigh, drawings 21 and 22, in "John Le Conte Lepidoptera of North America," Archives and Special Collections, University of Connecticut Library.

"More Small Ones"

In his old age, Abbot claimed he had drawn two to three thousand small drawings for LeConte. So far, we have accounted for approximately seven hundred of them. Where are the rest of them? They are hiding in plain sight in the midst of what was once thought to be John Eatton LeConte's own drawings.[48] Unlike the drawings of butterflies and moths that traveled to Paris and back, these other entomological drawings, the majority of which are of beetles, remained in the United States. They made the journey from Georgia to New York City and then moved to Philadelphia when the LeContes relocated to this vibrant center of natural history inquiry. There these drawings resided another half century until 1907, when they were acquired by the Missouri Botanical Garden in St. Louis. In 1953 the American Philosophical Society (APS) brought them back to Philadelphia, where they can be found in the library's extensive LeConte collection of archival materials.

The APS possesses thirty-seven hundred small drawings of insects that once belonged to Major LeConte.[49] In the 1870s these drawings had been organized into eight volumes, each volume carrying the title "Portfolio of Original Figures of Entomological and Other Subjects by Le Conte." Of these, hundreds if not thousands were drawn by Abbot.[50] Beetles take precedence in this collection of drawings: Volume 1 contains 654 figures of Coleoptera, volume 3 displays 698 figures of beetles, volume 5 has 657, and volume 4 combines 155 drawings of mostly weevils, with 222 drawings of wasps and bees. Many of these drawings are clearly by Abbot, who told Escher in 1813: "I am at present Employed in Drawing the

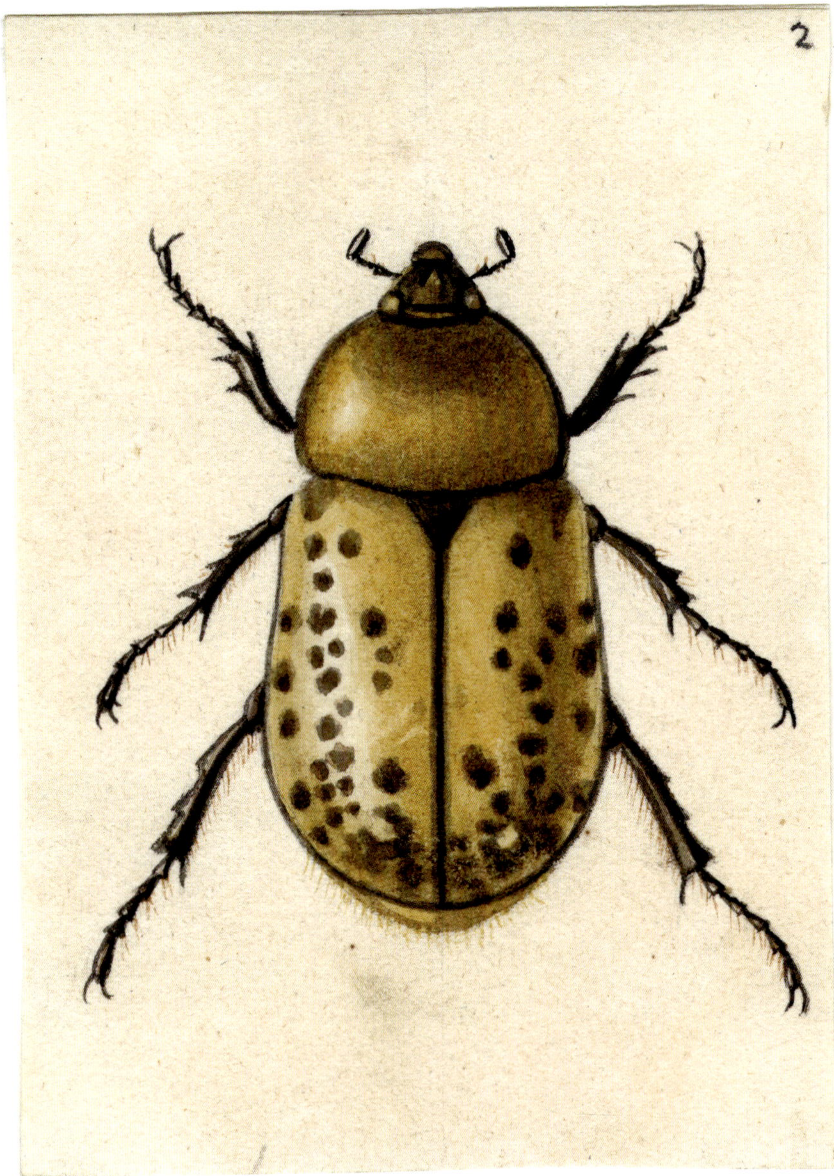

FIGURE 5.14.
Abbot, eastern Hercules beetle (*Dynastes tityus*), watercolor drawing, "Portfolio of Original Figures of Entomological and Other Subjects by Le Conte," vol. 5, p. 53, American Philosophical Society, Philadelphia.

FIGURE 5.15.
John Lawrence LeConte, beetles, graphite drawings, "Portfolio of Original Figures of Entomological and Other Subjects by Le Conte," vol. 3, p. 16, American Philosophical Society, Philadelphia.

Georgia Insects for a Mr Le Compte, particularly the Coleoptera, he is very anxious for me to go on as fast as my Time will permit."[51] Some of the beetles drawn by Abbot are quite lovely, particularly those that are similar to the beetle drawings that Francillon had collected. Most of Abbot's drawings in the APS collection, however, were made in the second and third decades of the nineteenth century, ten to twenty, even thirty, years after those drawn for Francillon.

Mixed in with Abbot's drawings of beetles are many drawn by LeConte and his son, John Lawrence LeConte, who became one of nineteenth-century America's most prominent entomologists. Both father and son loved beetles, and it seems likely that either of them could have produced the graphite drawings of the coleopterans, some of which were named by a LeConte (presumably

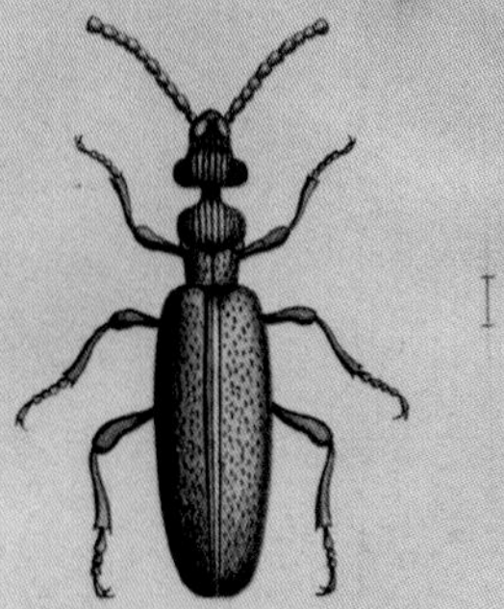

Dacoderus striaticeps
Colorado Desert. Lec.

Centrioptera muricata
Colorado Desert. Lec.

Cryptoglossa verrucosa
Colorado Desert. Lec.

Pl. 6 fig 5.

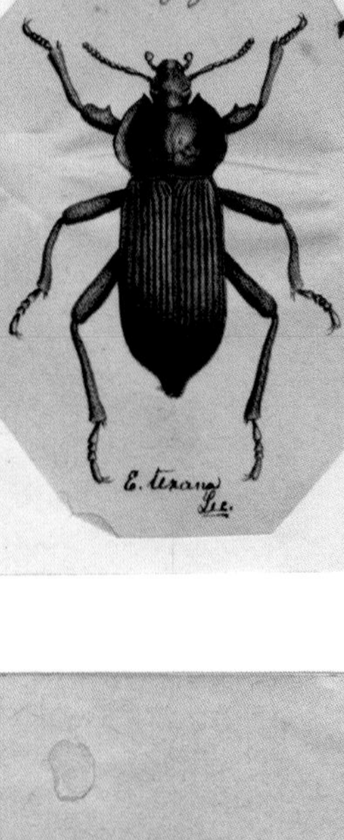

E. texana
Lec.

Eleodes armata
Colorado Desert. Lec.

Eleodes ventricosa
Texas Lec.

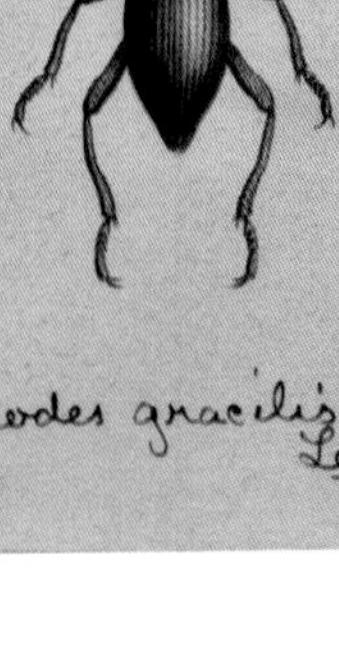

Eleodes gracilis
N. Mex. Lec.

Eleodes laticollis
S. Diego Lec.

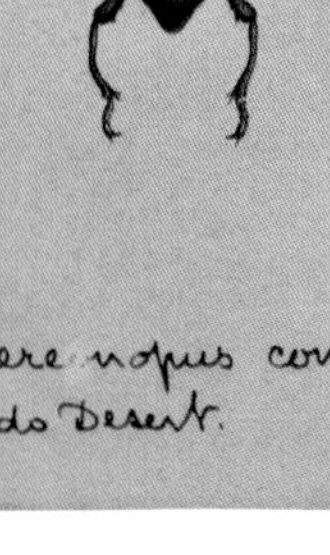

Cerenopus concolor
Colorado Desert. Lec.

John Lawrence). For instance, two drawings, one of "Eleodes armata Lec.," described as being from the "Colorado Desert," and "Eleodes laticollis Lec.," from San Diego, are exquisitely rendered entirely in graphite. The younger LeConte went on transcontinental collecting expeditions to the Rocky Mountains in 1844–46 and again in 1867, the Great Lakes in 1848, and California in 1849–51, these expeditions offering strong evidence that it was he who drew these beetles from the American West.[52]

The reason LeConte mixed Abbot's drawings with his own and his son's was to create a comprehensive insect collection consisting of drawings that he could then organize using Linnaean categories, arranging them by family, genus, and species. This paper museum could have been used to augment or even to substitute for a collection of actual specimens that would have required more care and space to maintain properly. What Abbot did for LeConte was to find, collect, and depict insects that were mostly unknown in the sense that they had not yet been described according to Linnaean systematics. The small drawings of beetles held in volumes 1, 3, 4, and 5 bear the marks of those who have tried to identify them. The majority of the Linnaean names scribbled on these drawings are in handwriting that is loose, slanted, and recognizable as Major LeConte's. His notations are supplemented in handwriting that most likely was his son's or could possibly have belonged to B. M. Everhart, who bought these entomological drawings in 1888 and owned them until 1905, or to an eager curator at the Missouri Botanical Garden during the drawings' sojourn there from 1907 to 1953.[53]

Typical of the kinds of annotations that the small beetle drawings wear is a penciled identification, "Geotrupes excavatus," written in a loose scrawl under a drawing of a beetle in volume 5 that certainly looks like Abbot's work (although I think the beetle might be a *Geotrupes blackburnii*). In some cases, the penciled identifications are overwritten in a fine pen in slanted handwriting, the pen's mark reaffirming what might have been an initial guess at the proper name for a new species. In other cases, there is the smaller, tighter, up-and-down handwriting of a pen-wielding taxonomist who has added names of species. Most of these small drawings with annotations have been trimmed to fit onto a page in a uniform manner, a process that has left some words trimmed as well, cutting "brunnicollis," for instance, down to "brunnico" and making it difficult to decipher what the taxonomist had written. The trimming strikes me as careless, as it tends to undermine the whole point of the endeavor: to construct a visual inventory of all North American Coleoptera, an impossible task, given the number of species there are of Coleoptera.

Who gets to describe and name a new species was what was at stake in this drive to collect as many insects as possible, and Abbot served this endeavor by collecting specimens and drawing them for LeConte, Harris, and

FIGURE 5.16.

Abbot, earth-boring scarab beetle (*Geotrupes* sp.), det. J. K. Wadleigh, watercolor drawing, "Portfolio of Original Figures of Entomological and Other Subjects by Le Conte," vol. 5, p. 39, American Philosophical Society, Philadelphia.

others who were eager to have the honor of naming new species. Discussing who could take credit for the naming of a species, Thaddeus W. Harris probed LeConte's motives for suggesting they exchange specimens with each other. LeConte had sent Harris specimens from the South collected by "a man" (who was certainly Abbot), and Harris sent LeConte insects he had collected in Massachusetts. Harris seemed to have concerns about the collaborative nature of this exchange and wanted to make sure that LeConte understood that "*the right of priority in imposing a specific name* does not depend upon this name having been given in a cabinet or a bare catalogue, but in its being designated to the world by a *description*, & the *first describer* alone, can claim this right."[54] Oscillating between collaboration with other naturalists and being competitive with them is typical behavior within taxonomic circles when what is at stake is the right to name a new species and the recognition that comes with that.

Most of the attempts to identify species in the APS's portfolios are confined primarily to the beetle drawings. The few annotations that appear on the drawings of Hymenoptera (bees and wasps) in volumes 4 and 6 are not in John Eatton LeConte's handwriting, nor does he seem to have annotated any of the 164 drawings of Diptera (flies) in volume 6 and the 233 dipteran drawings of volume 7. Likewise, no LeConte annotations appear on the 126 drawings of spiders in volume 8. These drawings of spiders, flies, bees, and wasps, all without LeConte's annotations, look very much like Abbot's work.[55] While Abbot's beetle drawings in general hold up to scrutiny, the Hymenoptera in volumes 4 and 6 and the Hemiptera in volumes 2 and 8 lack the crispness and finesse that his earlier work possesses. Some Diptera are downright fuzzy, suggesting these drawings that date from the late 1820s and early 1830s may be a product of the notorious "inferior copyist" (the aged Abbot, I presume), but even these display correctly those characteristics that a taxonomist would have needed to see to identify or name them.[56] Abbot's dipteran drawings in these volumes, while not executed with the precision of his earlier work, still possess scientific value. For instance, written in pencil on a drawing of a nonbiting midge in an unidentified hand is the family name "Chironomidae," which indicates that Abbot's drawings, even these somewhat fuzzy ones, contain enough information for identification to the family and genera levels.

The LeConte annotations on the beetle drawings reflect the LeContes' deep interest in Coleoptera. The elder LeConte asked Thaddeus Harris, with whom he hoped to exchange insect specimens, to send him coleopterans and lepidopterans only. Harris responded to LeConte's proposition:

> You have generously left the terms of exchange to me, & I have the assurance of my excellent young friend Dr Pickering that I can entirely depend upon

> you. Your proposition embraces only Coleoptera & Lepidoptera—I could have wished it included the other Orders also. However if my proposal is accepted by you, I will send you specimens of every species (in the orders of Coleoptera and Lepidoptera) which appear to be new, or of which I have doubt, & will also include some others which may be useful to you for the purpose of comparison, description, or to establish localities.[57]

LeConte's son, John Lawrence LeConte, whose field of specialization became Coleoptera, published a six-hundred-page monograph in 1883 identifying and naming eleven thousand North American beetle species.[58] The father and son's passion for Coleoptera is evident in their graphite drawings of beetles found in these portfolios, some of which sit side by side with Abbot's watercolor beetle drawings. Among the LeConte drawings are some truly stunning renditions, complete with dramatic handling of light, texture, and form (see figure 5.15).

Adding to the confusion caused by the mingling of the LeContes' and Abbot's artwork is what happened to John Eatton LeConte's enormous collection of drawings after his death in 1860. His son inherited his father's papers and drawings, including those by Abbot, all of which seem to have been in a state of disarray, the result perhaps of his father's "chronic invalidism," brought on by rheumatoid arthritis.[59] According to Asa Gray, LeConte senior was "a most untiring student" and "keen observer" of the natural world who had generated "much manuscript," which "his son has had mounted in albums suitable for inspection."[60] These must be the albums now in the American Philosophical Society's possession. Why they were sold at auction in 1884 after the younger LeConte's death is unclear.[61] A possible explanation as to why they were not given or sold to the Academy of Natural Sciences, an organization both LeContes were active members of, is that over the course of the two decades that separated the two men's deaths, the scientific value of these albums had waned. Asa Gray's opinion of LeConte senior's "much manuscript, the usefulness of which has been superseded by subsequent research," may have been shared by others, including the younger LeConte. He too may not have valued his father's collection of drawings, regarding them as a quaint remnant of what was regarded as old-fashioned natural history with its accumulation of drawings and specimens, methods that did not measure up to the rigors of "modern" taxonomy.[62] The younger LeConte and other professional entomologists of the late nineteenth century valued the verbal description over the visual representation of a specimen.[63] His six-hundred-page treatise on Coleoptera contains no illustrations, and in this book's introduction J. L. LeConte reasserts the hierarchy that privileges taxonomy over fieldwork: "The collection and the observation in the field of these small, but beautiful objects furnish a most agreeable and useful preliminary training to their investigation, but are in themselves, until subjected to the critical revision of the student, of small value for systematic or economic science, in so far as that they aid but little in forming the classification and stable nomenclature, upon which the knowledge of the objects treated of must rest, in order to permit them to be intelleigently spoken of."[64] Whatever John Lawrence may have thought of his father's work, when John Lawrence died twenty-five years later, his library was auctioned off along with his father's collection of drawings in 1884. Of the eleven LeConte portfolios described in the auction catalog, all but one can be accounted for today. The portfolio containing "161 colored drawings of Insects included in the family Neuroptera" (lacewings and antlions) seems to be missing.[65] John Eatton LeConte's own frog and turtle drawings found their way into the University of Georgia's Hargrett Library after a sojourn in the De Renne family library.[66] His drawings of native violets are housed in the University of Notre Dame's Hesburgh Library, and the entomological drawings—a combination of Abbot's drawings and those by the LeContes—are safely housed in the American Philosophical Society's library.[67]

Curation Versus Use

Of the major collectors of Abbot's drawings and notes, LeConte and Francillon stand out as having owned collections of well over two thousand drawings each. For LeConte, Abbot's drawings were part of the scientific process of bringing order to bear on the diversity and profusion of the natural world through Linnaean systematics. The thousands of little drawings that LeConte and Francillon bought from Abbot were, for LeConte, a means to an end; however, for Francillon, they were an end in and of themselves. LeConte, as a scientist, used his collection of drawings to think with, while Francillon, true to the habits of a collector, cared for his collection of Abbot's drawings as would a museum curator, regarding them as precious objects in need of protection from the forces of decay, disorder, and indifference. Francillon pasted Abbot's drawings onto sheets of fine wove paper. Affixed under these drawings are strips of paper carrying transcriptions of Abbot's notes. These sheets of paper bearing Abbot's images and words were bound in elegant and sturdy volumes that luckily survived the World War II bombing of London and are available today for inspection at the Natural History Museum's library, where they can still be called upon to offer insight into questions concerning the identity, habitat, population, biology, and ecology of North American arthropods, as well as to illuminate the material practices of the late eighteenth- and early nineteenth-century study of the natural world.

In my description of these two ways of collecting natural history drawings, my preference for Francillon's careful curation of Abbot's drawings and notes over what I have characterized as LeConte's less than perfect stewardship may be detectable. Less than perfect, I say, because when Abbot's drawings were combined with those drawn by the two LeContes under the rubric of cataloging and classifying the insect world, these material artifacts became unmoored from the context in which they were made. When Linnaean systematics were brought to bear on these objects, the origins and provenance of these drawings were obscured, things that those of us who study the visual and material culture of natural history want to know.

Where a drawing or a specimen ends up, in which archive, library, or museum it lands, shapes how we today can engage with it and what we can learn from its material state. In natural history museums, scientific practice tends to trump the concerns of art historians and historians of science. Kathy Day, former curator of Mollusca at London's Natural History Museum, explained this to me more than a decade ago when she took me down to the museum's basement to see Sir Joseph Banks's shell collection. The shells Banks had collected on Captain Cook's first voyage (1768–71) had been placed in carefully crafted, little tin boxes designed to fit into the drawers of Banks's mahogany cabinet. When Day opened the cabinet and pulled out a drawer, she apologized for the missing shells, as some of the boxes were empty. She explained that the museum's malacologists would every so often raid the cabinet, looking for a specimen that would aid them in whatever taxonomic quest they were on, and they would often forget to return it to the Banks collection or decide to put the shell in what they thought of as its proper place in terms of systematics. These malacologists did not care who had collected the shell but valued it for what it could tell them about a species.

Art museums, on the other hand, value provenance and keep detailed documentation on an object and all its past possessors. Libraries tend to be less interested in provenance and more concerned with the balancing act of keeping objects safe and sound and preserving them as part of our heritage while at the same time making these objects available to the widest possible audience. The Abbot/LeConte drawings held in the American Philosophical Society's library, for instance, underwent safekeeping measures in the 1980s when the original volumes, "portfolios," as LeConte named them, were dismantled. These small drawings were removed (or cut out) from the nineteenth-century paper and then affixed to modern archival paper, with several drawings arranged on each sheet. These sheets were then placed in mylar plastic sleeves that went into a binder. There are eight binders of drawings, which correspond to the eight volumes of drawings the APS had acquired from the Missouri Botanical Garden. James A. G. Rehn, curator of insects at Philadelphia's Academy of Natural Sciences, described the portfolios in 1954 as "eight bound volumes or scrapbooks of folio size, with leather backs, board covers, and pages of heavy paper, on which the individual illustrations have been mounted." Rehn's description of these "scrapbooks," the term also used by the auctioneer in 1884, conveys the way in which small drawings were pasted onto large sheets of paper: "The figures, which total 4,043, are chiefly in water color, each a separate unit; these usually arranged nine on a page, although they sometimes vary to as few as four to a sheet, owing to the size of the respective figures."[68] These volumes, as described by Rehn, are quite similar to the portfolio of LeConte's drawings of turtles and frogs, sold at the same 1884 auction and now in the Hargrett Library. This portfolio still possesses its original binding, boards, and paper, though it is barely hanging together at this point. This unrestored portfolio can tell us much about past material practices of knowledge production and the history of natural history, and for these reasons, it is very precious.

I also understand the importance of preservation and the differing demands of authenticity and longevity that museums, libraries, and even antiquarian bookshops must navigate. Antiquarian book dealers sometimes tidy up old, tattered volumes to make them more attractive to buyers. This is what happened to Abbot's drawings that are now held at the University of South Carolina's Thomas Cooper Library. H. P. Kraus, a prestigious firm that dealt in antiquarian books, bought these drawings at a 1963 Sotheby sale in London (the catalog is titled *Catalogue of a Fine Collection of Rare Books and Drawings, on Ornithology, Botany, Entomology, and Other Natural History, Voyages and Travel, The Property of a Lady*) and had them affixed to archival paper, matted, and cradled in elaborate archival cases.[69] These efforts by the Kraus firm and the library of the American Philosophical Society to care for Abbot's drawings are admirable. However, it is unfortunate that the original materials, including binding, boards, and paper, were not kept or at least photographed and recorded. Manuscript materials, including drawings, are vulnerable to wear and tear, damp and mold, disregard and disarray. Countering these forces requires money and energy, time and space, and a belief in the value of manuscript drawings and a commitment to protect them. Books, as opposed to manuscripts, fare a bit better in this struggle to persist in a changing world, and it is to Smith and Abbot's beautiful book that we now turn.

6

“Beauty and Accuracy”

ABBOT’S SCIENTIFIC CONTRIBUTION TO *INSECTS OF GEORGIA*

IN MY QUEST TO DETERMINE the scientific accuracy of Abbot's drawings, I coaxed a few professional entomologists into looking at some of his drawings of moths, butterflies, and beetles. Unaware of Abbot's work, these museum curators and academics were kind enough to take time out from their busy schedules to indulge my request. In chapter 1 I told the story of Professor John Pickering, a lepidopterist and founder of https://www.discoverlife.org/, expressing his delight in seeing his favorite moth in my digital photographs of Abbot's drawings. When I showed Abbot's drawings of beetles to Max Barclay, senior curator of Coleoptera at London's Natural History Museum, he was surprised and even pleased, saying that Abbot's drawings of beetles were very good. When I pressed him to explain further, he said, pointing to a figure of a beetle, "Hmm . . . those proportions are right." Professor Joe McHugh, also a coleopterist, confirmed that many of Abbot's drawings of beetles are "identifiable to the species level and most of the rest are recognizable to the generic level."[1] Jessica Wadleigh, collections assistant in the Division of Insects at the Field Museum in Chicago, was able to identify to the species level almost all of the insects and spiders in the drawings by Abbot that I gave her to examine. Professor James Porter, an ecologist, butterfly collector, and curator at the Georgia Natural History Museum, was thrilled to see among Abbot's illustrations a northern pearly-eye (*Lethe anthedon*), not named until 1936. His realization that Abbot was familiar with lepidopteran species long before they had been "discovered" and scientifically named mirrors Avinoff's reaction when he recognized in his collection of Abbot's drawings "the figure of Terias (Eurema) adamsi," which had been described, Avinoff said, "by my friend Lathy sometime about 1910, and here it was figured and remained unknown for 145 years!"[2] These responses suggest that Abbot's drawings, though products of their era, continue to work within scientific cultures of specialized sight while at the same time appealing to aesthetic sensibilities with their colorful detailing of marvelous insect bodies.

These anecdotes from my interactions with contemporary entomologists are evidence that Abbot's drawings are remarkable examples of what historian Lorraine Daston has called the Enlightenment's "peculiar economy of attention," her expression for highly trained, attentive, and selective ways of seeing.[3] In the process of depicting hundreds of previously undescribed and unnamed moths, beetles, and spiders, Abbot made decisions about which morphological features—shape of antennae, wing venation patterns, and the number of spurs on tarsi, for instance—should be represented to enable the specimen's identification and classification. Attending to the anatomical features most crucial to taxonomic identification, he rendered in two dimensions the intricacy and beauty of an insect's body. Abbot's entomological portraits owe their impact to Linnaean ways of seeing, with their emphasis on the visible forms of anatomy unaided by dissection and a microscope. These "aggressively selective" ways of looking at and depicting specimens give to his best entomological watercolors a quiet simplicity that lets the wondrous intricacy of the insect's body emerge under an attentive and knowledgeable gaze.[4] As such, they are an extraordinary contribution to the study of the natural history of North America.

James Edward Smith praised Abbot's drawings for *Insects of Georgia* (1797) for their "beauty and accuracy."[5] The author of an anonymous review waxed rhapsodic over Smith and Abbot's book: "The produce of judicious researches[,] . . . the volumes before us are of the highest value. What is given here are not fragments of knowledge, but the result of a series of finished observations. If the work allure and delight by splendour of appearance, and uniform elegance of execution, it still more surprises and instructs by the richness and novelty of its contents, the lucid order with which they are digested, the precision and vivacity of the designs, the modesty of method, and spirit of philosophy, that pervade the whole."[6] A friend of Smith's, William Jones, an amateur lepidopterist and gifted illustrator of butterflies, congratulated him on the publication of *Insects of Georgia*, writing somewhat ironically: "I took it under my arm, but soon found it sufficiently weighty . . . so I laboured abundantly with my load, in expectation of high gratification from the contents when I should get home—and truly I was not disappointed. . . . [U]pon the whole, it has three great requisites to a modern publication—good letter, good paper, and showy plates."[7] William Swainson described "this noble work" as "unquestionably one of the most beautiful and valuable that this or any other country can boast of" and one of the "best illustrative publications upon Insects that have ever been given to the world."[8] Adrian Haworth's delight with Smith and Abbot's book bears repeating here, his 1807 review declaring it "a magnificent work in folio," "the whole Plants as well as Insects being scientifically delineated and described, that this publication is to the full as valuable

Papilio Ilioneus.

to the Botanist, as it is to the Entomologist: we never before beheld the sister sciences walk so closely, and so engagingly hand in hand, as in this interesting volume.—it is truly a Flora et Entomologia."[9]

This chapter argues that *Insects of Georgia* is a remarkable achievement because it is not only a successful marriage of art and science but also the product of a collaboration that bridged two different ways of studying the insect world. Abbot's methods were based on careful observation, visual representation, and verbal description of living creatures he had reared. While Smith's methods were also observation based, he conducted the work of identifying and classifying specimens in museums and libraries far from the field where Abbot did his work. Abbot and Smith were fortunate that their collaboration occurred in the 1790s before the boundaries between practical naturalists and systematists had hardened as the study of insects became the specialized discipline of entomology.[10]

Earlier in the eighteenth century, those who loved insects and studied them consisted of various disparate groups, among them natural history collectors, professional scientists, avid amateur naturalists, skilled specimen hunters, and poets and artists intrigued by the beauty and intricacy of insect bodies. Initially, members of these different groups were gratified to find others who loved insects and were grateful for the comradery that these shared interests generated. However, over time tensions arose as insect enthusiasts challenged each other over whose approach to the study of insects was best and whose knowledge counted more. With the rise of entomology as a specialized field within zoological studies, observational approaches to the study of insects like those Abbot employed were perceived by more established systematic entomologists as less scientifically rigorous than their own more museum-based and text-oriented work of classification. The distinction between "learned" or systematic naturalists and "practical" naturalists who were field based, along with its implied hierarchy, rankled many naturalists who felt underappreciated by those who regarded taxonomy as the core mission of the life sciences.[11]

In this chapter I place Abbot's scientific contribution to *Insects of Georgia* within the context of the emergence of entomology in the early nineteenth century as a scientific discipline. Before diving into this fraught world of rivalry between various communities of insect lovers, I analyze Smith's and Abbot's differing contributions to the text of *Insects of Georgia* and explore the implicit hierarchy at work that placed Smith's more theoretical musings over Abbot's more grounded empirical knowledge acquired through experience of living and working in the field. I then explore this tension between taxonomy and fieldwork that surfaces occasionally in *Insects of Georgia* more fully in a few heated exchanges within early nineteenth-century entomological circles. My aim here is to provide a framework that might help explain the waning of Abbot's scientific

FIGURE 6.1.
Abbot, "Sassafras Black Swallow-tail Butterfly, *Papilio Ileoneus*," spicebush swallowtail (*Papilio troilus*), engraving, TAB. II, Smith and Abbot, *Insects of Georgia* (1797), Hargrett Rare Book and Manuscript Library, University of Georgia Libraries.

reputation in the latter half of the nineteenth century. As a whole, this chapter and the next argue that Abbot's scientific contribution to *Insects of Georgia* is far more significant than has been generally recognized and that it is time to acknowledge his rightful place in the history of American natural history.[12]

Art and Science

Not simply a division of labor, with Abbot acting as the artist and Smith as the scientist, the collaboration between Smith and Abbot was quite complex. The illustrations for this book contain scientific information that could be used to identify and classify not only the pictured lepidopterans but also the plants on which their larvae fed. Abbot's illustrations are based on firsthand knowledge of the life histories of the butterflies and moths he had raised, as Smith explains in his preface:

> The materials of the following work have been collected on the spot by a faithful observer, Mr. John Abbot, many years resident in Georgia, who, after having previously studied the metamorphoses of English insects, pursued his enquiries among those of Georgia and the neighbouring parts of North America. The result of his observations he had delineated in a style of beauty and accuracy which can scarcely be excelled, and has accompanied his figures with an account as well as a representation, of the plants on which each insect chiefly feeds, together with many circumstances of its manners, times of the different metamorphoses, and other interesting particulars. For all such facts recorded in these pages the public are entirely obliged to Mr. Abbot.[13]

Because Abbot had, according to Swainson, "contrived to breed nearly the whole of the Lepidoptera," he knew what plants these caterpillars ate and included them in his drawings along with their caterpillars and pupae.[14] *Insects of Georgia*, as Haworth noted, is a flora et entomologia, but it is much more than that, since it portrays metamorphosis and the various stages Lepidoptera go through in this process.[15] The pretty flowers and snipped leafy twigs in these illustrations are not simply there for added visual appeal or to pique botanical interest; they are included to document their role in the metamorphic changes each lepidopteran undergoes.

Caterpillars are featured on each of the 104 plates of *Insects of Georgia* not just because they are visually appealing or mildly amusing but also because they could be called upon to help in determining the species of the depicted butterfly or moth. These detailed drawings provided lepidopterists and taxonomists with morphological information about the caterpillars that could help them to distinguish one species from another. Smith often turned to Abbot's drawings of caterpillars to aid him in ascertaining species of the book's Lepidoptera. Because of Abbot's deep familiarity with these caterpillars, having raised all those depicted in the 104 plates, Smith knew he could trust him to depict the caterpillars and the adults with precision. When Smith compared Abbot's drawings with published illustrations and descriptions of the same species, he sided with Abbot's rendition, insisting that "we must rely on the accuracy of Mr. Abbot."[16]

Abbot used his skills as a watercolorist and his observations as a naturalist to attend to the often odd bodies of caterpillars, rendering in great detail their stripes, dots, fuzz, tufts, feet, horns, tails, and other protuberances, such as those displayed by the caterpillars of the "Kitten Moth" (*Furcula borealis*): "When disturbed it shoots out of the ends of its forked tail two soft orange-coloured threads."[17] Abbot provides two figures of this caterpillar, each representing a different instar, the stages of a caterpillar's growth as it molts and sheds its too-tight skin. For strange tufts of hair, there is nothing quite like the caterpillar of the "Waved Yellow Egger Moth, *Phalaena Opercularis*" (now *Megalopyge opercularis*), with its masses of hair tumbling down its back: "The caterpillar is shaped like the roof of a house, the hairs rising on each side gradually to a ridge down the middle of its back."[18] Lepidopterist Donald W. Hall of the University of Florida notes that the "puss caterpillar" of the southern flannel moth, as this eggar moth is known today, becomes "progressively more 'hairy' with each molt."[19] Abbot's illustration captures this hairy creature's distinctively arched back, as well as the cocoon's unusual shape, with an end that is flat and "opens and shuts like a door, and is fitted with the greatest exactness."[20] (See figure 6.3.)

In his deliberations over distinguishing species that closely resembled one another, Smith often referred to the caterpillars Abbot had drawn, focusing on their colorful patterns of stripes and spots as a way to register difference among species. For instance, in his attempt to distinguish the "Black-Barred Swallowtail Butterfly, *Papilio Ajax*" (*Eurytides marcellus*, zebra swallowtail), from its European look-alike, "*Papilio Podalirius*" (*Iphiclides podalirius*, scarce swallowtail), he turned to their larvae to help make the case for them being different species. Arguing that these two kinds of swallowtails are distinct species, Smith wrote: "In the caterpillar state these two insects differ considerably, that of the *Podalirius* not having either the horns of ours, nor the transverse black, blue, and yellow stripe."[21] The colors and patterns of these caterpillars were key to his argument that these European and North American swallowtails were different species. (See figure 6.5.)

Earlier depictions of North American Lepidoptera, such as those in Drury's *Exotic Insects*, do not contain images of caterpillars.[22] This is simply because Drury may not have had many caterpillars as specimens in his collection of

FIGURE 6.2.

Abbot, "Kitten Moth, *Phalaena Furcula*," white furcula (*Furcula borealis*), original drawing, TAB. LXXI, Smith and Abbot, *Insects of Georgia* (1797). Courtesy of John Work Garrett Library, Sheridan Libraries, Johns Hopkins University.

FIGURE 6.3.
Abbot, "Waved Yellow Egger Moth, *Phalaena Opercularis*," southern flannel moth (*Megalopyge opercularis*), original drawing, TAB. LIII, Smith and Abbot, *Insects of Georgia* (1797). Courtesy of John Work Garrett Library, Sheridan Libraries, Johns Hopkins University.

FIGURE 6.4.
Abbot, "Waved Yellow Egger Moth, *Phalaena Opercularis*," southern flannel moth (*Megaloppyge opercularis*), original drawing, TAB. LIII (cropped), Smith and Abbot, *Insects of Georgia* (1797). Courtesy of John Work Garrett Library, Sheridan Libraries, Johns Hopkins University.

FIGURE 6.5.
Abbot, "Black-Barred Swallow-tail Butterfly, *Papilio Ajax*," zebra swallowtail (*Eurytides marcellus*), original drawing, TAB. IV, Smith and Abbot, *Insects of Georgia* (1797). Courtesy of John Work Garrett Library, Sheridan Libraries, Johns Hopkins University.

exotic insects. As discussed in chapter 2, Drury relied on agents who, as travelers or sojourners, did not raise butterflies and moths in foreign regions and instead employed locals, including children, servants, enslaved people, and Indigenous people, to catch the adult lepidopterans. Drury complained that his suppliers in North America did not know how to raise Lepidoptera, and even if they had had access to caterpillars, they would not have had the skills to taxidermy them. James Duncan complains in his 1841 publication, *The Natural History of Exotic Moths*, that "our acquaintance with the caterpillars of exotic moths is still so limited." He contends that "it is not difficult to account" for this problem, given the fact that the structural problems that collectors such as Drury faced continued well into the nineteenth century. Referring to non-European caterpillars, Duncan observed:

> It is by no means an easy matter to preserve them, at least in such a way as to retain their forms and colours in any degree of perfection. The greater number of individuals who take an interest in the subject are, moreover, only casual visitors to the countries where they make their collections, and they have neither time nor opportunity to rear the caterpillars till they disclose their moths; and unless this step be taken, the possession of the former is of comparatively little value. To this may be added, that the caterpillars are often difficult to discover, owing to the nature of their haunts, many of them living in the interior of trees, within the substance of herbaceous vegetables, in fruits, &c. But the want of this knowledge, from whatsoever causes it may arise, has hitherto been a principal cause, and will probably long continue to be so.[23]

In 1774 Drury wrote to Abbot just after he had arrived in Virginia, approving of his plan to raise Lepidoptera: "I am confident when you set about breeding you will find a numerous tribe of Lepidoptera that are unknown to us.—The truth is Nobody here knows what that Country produces because Nobody has ever tried to obtain Insects by that method. Therefore I have no doubt but you will bring a great many new Species to light."[24] Drury's support of Abbot's decision to go to North America stemmed from knowing he would benefit from Abbot's proficiency as a breeder of Lepidoptera.

Smith argued that Abbot's illustrations and accompanying notes, in providing important information about the stages of lepidopteran metamorphosis, were tackling larger philosophical questions about the natural world. Smith was critical of books such as Drury's *Exotic Insects* for not depicting the pupal and larval forms of these insects, even though the illustrations in Drury's book "display the complete insect in a degree of perfection that leaves scarcely anything to be desired." Smith complained that the illustrations in

FIGURE 6.6. Moses Harris, plate XXV, engraving, in Dru Drury, *Illustrations of Natural History, wherein are exhibited upwards of two hundred and forty figures of Exotic Insects*, vol. 1 (1770), HathiTrust Digital Library / University of Michigan.

Exotic Insects lacked important "philosophical" information: "Where are the metamorphoses through which these finished forms have passed? [W]here are their various habitations, foods, and manners?"[25] Smith even criticized the great entomologist and taxonomist Christian Fabricius for relying on collectors' specimens of Lepidoptera and failing to take the pupal and larval forms of these species into account in his masterwork on insect taxonomy, *Systema Entomologiae* (1775): "Fabricius . . . was not particularly acquainted with their metamorphoses, knowing them chiefly in cabinets, and therefore wanting [lacking] the great clue to a natural arrangement."[26] According to Smith, studying dead specimens instead of the living insects had led to an imperfect understanding of these organisms and caused confusion in identifying and classifying species. For a book on Lepidoptera to be of use scientifically, the stages of metamorphosis needed to be represented.[27]

Collaboration

Abbot's scientific contribution to this book was not confined to its illustrations. The written text of *Insects in Georgia* was coauthored, with Smith handling questions concerning taxonomy and Abbot providing natural history descriptions of the pictured lepidopterans' life history, behavior, seasonal distribution, geographic location, and, most importantly, what plants those hungry caterpillars ate.[28] Abbot's deep familiarity with the caterpillars he drew permeates his descriptions. Note, for example, the detailed description of the puss caterpillar's pupa, the hairy caterpillar of the "Waved Yellow Egger Moth, *Phalaena Opercularis*," discussed above: "This species always shapes its web or cocoon as in the figure, fastening it to a twig. The flat end opens and shuts like a door, and is fitted with the greatest exactness. The insect continues in this web all winter in the worm state, not changing to a chrysalis till within a short time of its final transformation into a fly."[29] Abbot most likely gleaned such details from his breeding journals, similar to those kept by Francillon and Moses Harris, in which they recorded data about when and where they found a caterpillar and on what plant, when the caterpillar entered the pupal stage, and when it emerged as a winged adult. The format of Abbot's descriptive passages, though, as Ronald S. Wilkinson, a bibliographer at the Library of Congress and author of several well-researched articles on Abbot, suggests, initially modeled on John Berkenhout's *Outlines of the Natural History of Great Britain and Ireland* (1769–72), departs from Berkenhout's spare prose with a focus on lepidopteran life history, including details about caterpillar behavior.[30]

With Abbot responsible for natural history description and Smith for classification, Smith's preface explains how this collaboration was translated onto the page: "His memorandums, not methodized by himself for publication, have merely been digested into some sort of style and order by the editor, who has generally added remarks of his own, in a separate paragraph and different type from the rest; and who has entirely to answer for the systematic names and definitions; that department having been left altogether unattempted by Mr. Abbot."[31] A printer's device (a short but shapely line) distinguishes Abbot's natural history descriptions from Smith's taxonomic discussions by separating the two authors' contributions (see figure 6.7). Smith, referring to himself as an editor, makes it clear that Abbot was responsible for the "facts" concerning the life cycles of the pictured lepidopterans. Smith's description of his role in this collaboration as an editor is surprisingly modest, given that he was responsible for the identification and classification of the plants and lepidopterans depicted by Abbot.

Before describing the different approaches that Abbot and Smith took to their writing tasks, a few more words about the nature of their collaboration are necessary. Some nineteenth- and early twentieth-century commentators have claimed that Abbot had no idea that this book, based on his work, was going to be published in London and that he learned of the book's existence decades later. According to entomologist Robert Percy Dow, who was active at the turn of the twentieth century, "Smith never saw or heard directly from Abbot. Abbot knew nothing of the book until long after publication."[32] We cannot know whether Abbot and Smith wrote letters to each other, since most of Abbot's correspondence is long lost. No letter from Abbot has been found in Smith's voluminous correspondence, which lends some credence to Dow's assertion. However, in among Smith's papers is Abbot's manuscript for *Insects of Georgia*, which Abbot had titled "A Natural History of North American Insects, Particularly those of the State of Georgia. Including the changes of the principal Insects of those parts, together with the plant or flower each species feeds on, in the Natural Colours. Drawn from Nature by John Abbot many years Resident in those parts. With Notes Scientific and Illustrative" (no date [prior to 1797]). This manuscript contains introductory remarks addressed to someone to whom Abbot gave permission to edit the manuscript for stylistic correctness, contribute morphological descriptions, and name unnamed species:

> As I intended the following, I think you may still publish it as a separate Work from any other you are at present engaged in. However if you think otherwise you may only mention my Name now & then. As we are all naturally fond as my Lord Bolingbroke observes of recording and immortalizing our great Works.
>
> You may therefore prune and trim what you please of the following rude Notes. I shall therefore not marshall them in any Order, take them as they occur.
>
> I have not pretended to describe them in any scientific manner, leaving that for you[r] superior Abilities; nor yet their colours, but only what you cou'd not be informed of otherwise.[33]

Abbot is clearly very aware that someone who possessed skills as a writer and a taxonomist was going to do the work of getting these volumes ready for publication. Although he concedes authority to this editor over systematic nomenclature and prose style, the first sentence of the above statement implies something altogether different: He had "intended the following" manuscript to be "a separate Work from any other you are at present engaged in." This suggests that Abbot hoped he would be the recognized author of this volume, with the unnamed editor as a silent partner. But in the next sentence he quickly relinquishes this possibility and gently pokes fun at his own desire for immorality through "our great Works." Yet he manages to sound a poignant note over the fear that his "Name" would not be linked with this book, obscured by that of a collaborator, which is indeed what happened.

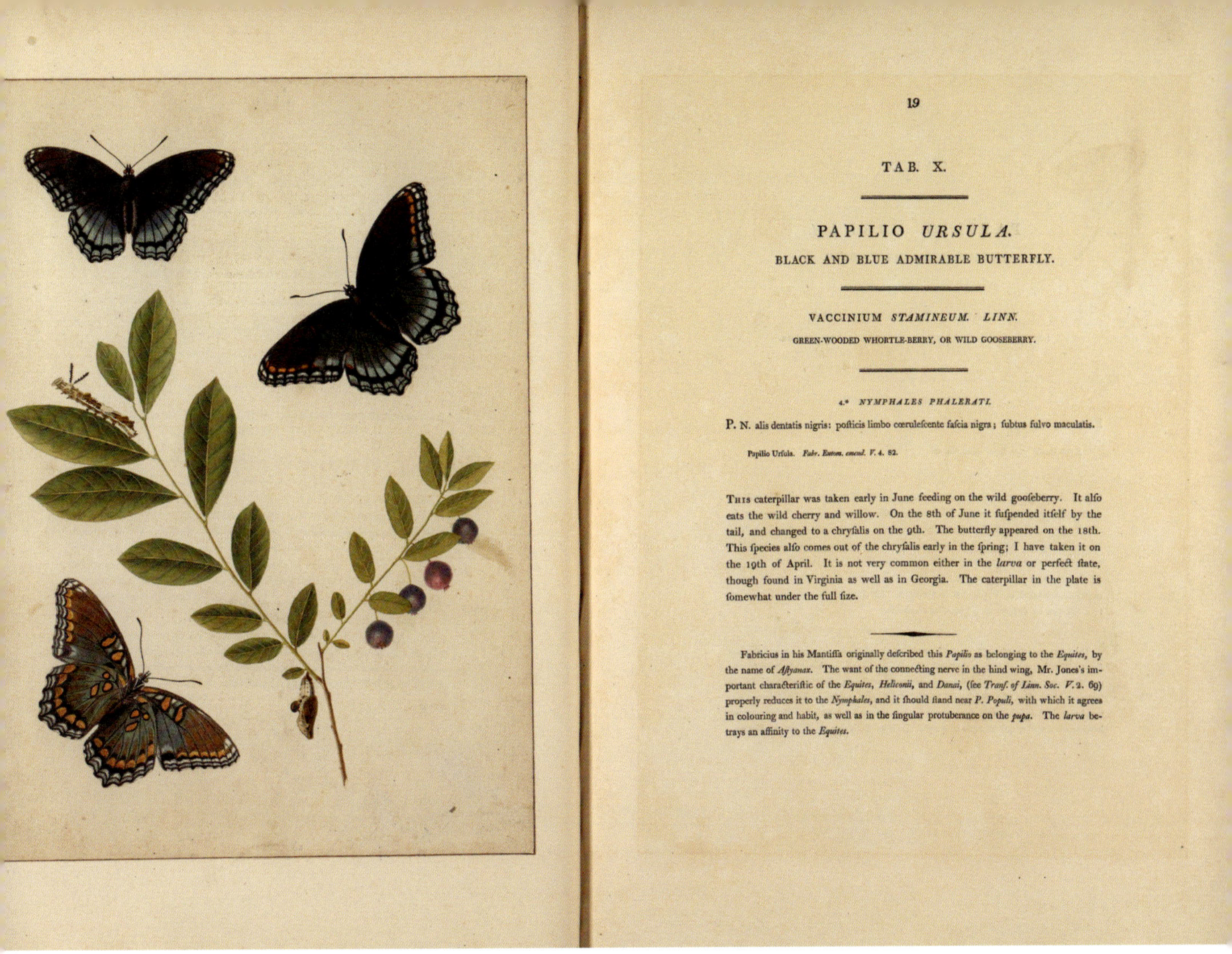
19

TAB. X.

PAPILIO *URSULA.*

BLACK AND BLUE ADMIRABLE BUTTERFLY.

VACCINIUM *STAMINEUM. LINN.*

GREEN-WOODED WHORTLE-BERRY, OR WILD GOOSEBERRY.

4.* *NYMPHALES PHALERATI.*

P. N. alis dentatis nigris: pofticis limbo cœrulefcente fafcia nigra; fubtus fulvo maculatis.

Papilio Urfula. *Fabr. Entom. emend. V.* 4. 82.

This caterpillar was taken early in June feeding on the wild goofeberry. It alfo eats the wild cherry and willow. On the 8th of June it fufpended itfelf by the tail, and changed to a chryfalis on the 9th. The butterfly appeared on the 18th. This fpecies alfo comes out of the chryfalis early in the fpring; I have taken it on the 19th of April. It is not very common either in the *larva* or perfect ftate, though found in Virginia as well as in Georgia. The caterpillar in the plate is fomewhat under the full fize.

Fabricius in his Mantiffa originally defcribed this *Papilio* as belonging to the *Equites*, by the name of *Aftyanax*. The want of the connecting nerve in the hind wing, Mr. Jones's important characteriftic of the *Equites*, *Heliconii*, and *Danai*, (fee *Tranf. of Linn. Soc. V.* 2. 69) properly reduces it to the *Nymphales*, and it fhould ftand near *P. Populi*, with which it agrees in colouring and habit, as well as in the fingular protuberance on the *pupa*. The *larva* betrays an affinity to the *Equites*.

FIGURE 6.7. Abbot, "Black and Blue Admirable Butterfly, *Papilio Ursula*," red-spotted purple (*Limenitis arthemis astyanax*), original drawing, TAB. X, Smith and Abbot, *Insects of Georgia* (1797). Courtesy of John Work Garrett Library, Sheridan Libraries, Johns Hopkins University.

Whom Abbot was addressing in these "Introductory Notes" is still a mystery. Francillon is a candidate, since he was capable of describing Lepidoptera in a "scientific manner," and he was both Abbot's agent and his friend, which would account for the ironic, self-deprecating tone of Abbot's introductory remarks.[34] Another theory is that Francillon may have sold Abbot's drawings to the publisher James Edwards, who then asked Smith to take on the role of editor and taxonomist.[35] Smith, an experienced writer, taxonomist, and botanist, fits Abbot's description of "you" better than Edwards, who was not a naturalist, or even Francillon, who, though a knowledgeable insect collector, had published very little.[36] Either way, the manuscript ended up in Smith's possession, sent to him by Abbot or passed along to Smith by an intermediary, possibly Francillon or Edwards. Although we cannot establish that Abbot communicated directly with Smith about this project, Smith's reworking of Abbot's prose and identification of the Lepidoptera Abbot drew constitute, I believe, collaboration, and a very successful one at that.

Abbot's manuscript's descriptions of the life histories of the pictured moths and butterflies resemble the notes that he routinely sent along with his drawings to his clients, and, as expected, these notes are filled with information about

FIGURE 6.8.
Abbot, "Large American Tiger Moth, *Phalaena Virgo*," virgin tiger moth (*Apantesis virgo*), original drawing, TAB. LXII, Smith and Abbot, *Insects of Georgia* (1797). Courtesy of John Work Garrett Library, Sheridan Libraries, Johns Hopkins University.

metamorphosis, caterpillars, and the plants upon which they feed. Smith edited Abbot's descriptive passages, mostly regularizing spelling and grammar and turning Abbot's fragments into complete sentences.[37] The passage describing a "Black and Blue Admirable Butterfly" (today known as the red-spotted purple butterfly, *Limenitis arthemis astyanax*) is typical of Abbot's contribution (with minor edits by Smith) to *Insects of Georgia*:

> This caterpillar was taken early in June feeding on the wild gooseberry. It also eats the wild cherry and willow. On the 8th of June it suspended itself by the tail, and changed to a chrysalis on the 9th. The butterfly appeared on the 18th. This species also comes out of the chrysalis early in the spring; I have taken it on the 19th of April. It is not very common either in the larva or perfect state, though found in Virginia as well as in Georgia. The caterpillar in the plate is somewhat under the full size.[38]

Beneath Abbot's description of the admirable butterfly is Smith's discussion of how he tried to identify it using printed sources such as Fabricius's *Philosophia Entomologica* (1778) and an article by William Jones: "Fabricius in his Mantissa [*Mantissa Insectorum* (1787)] originally described this *Papilio* as belonging to the *Equites*, by the name of *Astyanax*. The want of the connecting nerve in the hind wing, Mr. Jones's important characteristic of the *Equites*, *Heliconii*, and *Danai*, (see *Trans. of. Linn. Soc.* V. 2, 69) properly reduces it to the *Nymphales*, and it should stand near *P. Populi*, with which it agrees in colouring and habit, as well as in the singular protuberance on the *pupa*. The *larva* betrays an affinity to the *Equites*."[39] A printer's device separates the top passage, written by Abbot, from the bottom passage, written by Smith, making it clear to readers who wrote what. While Abbot was responsible for describing lepidopteran life histories, Smith was responsible for identifying and classifying the plants and the insects and giving binomial names in Latin to those plants and insects that Abbot had identified by their common or local name. Smith also provided the systematic description of each insect in Latin and a list of the authors, if any, that had described the pictured insect.[40]

A comparison of Smith's and Abbot's entries reveals two very different modes of doing natural history description. Abbot's experience of catching and raising butterflies and moths permeates his description of the volume's butterflies and moths. He notes, for instance, that the "Brown and White-Margined Moth" "may frequently be found sitting on the ground in foot-paths in the spring," and the "Blue-spot Ringlet Butterfly" "frequents the sides of rivulets, or *branches*, as they are called in America," details that convey how he might have initially encountered these butterflies.[41] In contrast to Abbot's interest in insect behavior, Smith's description of the "Blue-spot Ringlet Butterfly" is focused on his strug-

gles to find a reference to this butterfly in the scientific literature: "We have not been able to detect this fly in any systematic writer, nor is there a figure like it in Cramer. It exists however among Mr. Jones's valuable drawings, but has not till now received a name." Rejecting the idea that it is the same species as Fabricius's *Phocioni*, Smith decided to name this species "Papilio Areolatus."[42]

The work of identifying these species, which was Smith's contribution, involved examining all the available publications that described or pictured the species under question. This kind of work is text based and dependent on access to fine libraries and museums with large collections of specimens that could be consulted in the process of identification and classification. Smith's search for the correct scientific name for the "Large American Tiger Moth," for instance, led to frustration with his sources: "Clerck's figure of this insect is very bad, but it can be intended for no other. He represents the abdomen red above and black below; and Linnaeus seems to have described it so from his figure, for that description is not applicable to our species, nor any one of those nearest allied to it."[43] Abbot's work, on the other hand, involved caring for the actual living creatures, which was not always an easy task. Of the "Large American Tiger Moth," Abbot writes somewhat poignantly: "The caterpillar when kept in confinement is apt to die before it changes to a chrysalis."[44] Abbot also found it difficult to raise hawkmoth larvae, which bury themselves in the dirt and undergo metamorphoses underground: "Of those insects that go into the earth, and breed twice in a year, it is best to procure the spring caterpillars, which are much most likely to survive, the autumnal ones commonly dying in chrysalis in the winter."[45] (See figure 6.8.)

While Abbot faced problems of caterpillars who refused to eat in captivity or died when undergoing metamorphosis, Smith had his own difficulties to contend with as he wrestled with the published authorities, ranging from Linnaeus and Fabricius to Drury and Pieter Cramer (1721–76), a natural history collector who was an early adopter of Linnaean systematics, to arrive at the right name for these lepidopterans. Smith is quite frank about how he arrived at "systematic names and definitions" and openly discusses the process by which he named a new species or changed the name of an old one. An example of how Smith thought about the principles underlying the process of naming a species can be found in his commentary on the "Fringe-Tree Hawk-Moth" and "Gall-Berry Hawk-Moth." Smith criticizes the practice of bestowing a scientific name based on what he regarded as capricious and fanciful reasons, believing instead that moths needed names that would associate them with their food sources, thus supplying more information that would aid in their identification.

Smith defends, for instance, changing the names of two hawkmoths he felt were poorly named by Cramer. In his commentary on one of these moths,

FIGURE 6.9.
Abbot, "Fringe-Tree Hawk-Moth, *Sphinx Chionanthi*," rustic sphinx (*Manduca rustica*), original drawing, TAB. XXXIV, Smith and Abbot, *Insects of Georgia* (1797). Courtesy of John Work Garrett Library, Sheridan Libraries, Johns Hopkins University.

Smith acknowledges that renaming this species violated the Code of Zoological Nomenclature, which gives precedence to the name published first:

> Although every judicious writer ought to submit to the wise law, which by common consent has obtained among systematic naturalists, that no tolerably good trivial name, when once printed, should be changed by any following author, there being no end to altering names for the better, and the inconvenience of such changes, for the most part, more than counter-balancing their benefits: nevertheless we do venture at present to transgress this law. We do so on the authority of an opinion, if not decree, of *Linnaeus*, (Syst. Nat. 767, note) and also on the still higher authority of reason and sense. *Linnaeus* says, in the place just referred to, that trivial names [the second name of the binomial] of Insects taken from the plants on which they depend for subsistence, whenever such can be obtained, are preferable to all others, and ought solicitously to be sought for, in order that the dependence of the animal and vegetable kingdoms on each other may be closely kept in view.[46]

Smith's citing of Linnaeus's directive that naming a zoological species should take into account the relationship that species has with the botanical world may surprise modern-day critics of Linnaeus who characterize his system as devoid of interest in ecology and environment.[47] Smith goes on to argue with Cramer, the previous authority on this species, to say that "surely Cramer's inexpressive name of *rustica* for this noble insect, must give way to one descriptive of its food." Smith concludes that this moth should be named "*Sphinx Chionanthi*," after the fringetree, which he argues belongs to "the same natural order of plants as the Lilac and Privet, which though not native of its own country, this caterpillar can, according to Mr. Abbot's observations, occasionally eat."[48] Known today as the laurel sphinx (*Sphinx kalmiae*), the caterpillars of this moth do indeed dine on the white fringetree (*Chionanthus virginicus*), as well as on laurel, lilac, privet, ash, poplar, and honeysuckle.

Smith criticizes Cramer's classification of another hawkmoth. The "Gall-Berry Hawk-Moth," he argues, is not, as Cramer thought, the same as the "Fringe-Tree Hawk-Moth." Smith explains how he went about choosing its new name, "*Sphinx Prini*," naming it after the moth's host plant, "*Prinos Glaber*. Linn.":

> The reasons we have offered for changing the name of the preceding species must form our apology in the present instance. Hylarus, the name of a dog, has no natural nor fanciful connections with this insect, and could be tolerated only till we became acquainted with its natural food. Notwithstanding the great difference of size, our *S. Prini* has great affinity with the

FIGURE 6.10.
Abbot, "Gall Berry Hawk-Moth, *Sphinx Prini*," pawpaw sphinx (*Dolba hyloeus*), original drawing, TAB. XXXV, Smith and Abbot, *Insects of Georgia* (1797). Courtesy of John Work Garrett Library, Sheridan Libraries, Johns Hopkins University.

> *S. Chionanthi*, insomuch that no description can discriminate their upper wings. The want of yellow spots on the abdomen of that now before us makes their essential difference, to which might be added that the head of this is of an uniform brown, not white on the summit, and its thorax much less variegated than in the other. In a systematic arrangement these two species and *S. Carolina* ought certainly to stand together, and it is a matter of wonder how so good a writer as Fabricius should have placed them so very far asunder.
>
> If the figure of Cramer above quoted be intended for this moth, it is extremely inaccurate, and gives scarcely any idea of it; nor can we believe that author to have fallen so far short of the usual merit of his work. He must mean a different insect.[49]

Smith recognizes that not all naturalists will appreciate his careful reworking of "systematic names," saying to some that his "labours will appear trifling and fruitless," but those naturalists not interested in taxonomic subtleties he dismisses as caring "little about the philosophy of *things*."[50] For Smith, nomenclature should reflect natural affinities and should avoid the idiosyncratic, the personal, and the arbitrary in the act of naming a species. Naming a new species of moth after its host plant, for instance, would provide information about the species that would aid in identification.

Despite the hauteur that seeps into Smith's prose as he jostles with the era's great taxonomists and entomologists, his commentary manages to convey respect for Abbot's judgment concerning not just the flora the species feeds on but also the characteristics his drawings of Lepidoptera display. To determine the species to which a butterfly or moth belongs, Smith needed to pay attention to particular parts of the butterfly's body, such as the shape of its antennae, the markings on its abdomen, and the colorful pattern on its wings, and to examine the caterpillar's appearance and the shape of its chrysalis. He relied on Abbot's drawings for this information. Smith used Abbot's drawing of the "Great Yellow-Underwing Moth," for instance, to argue that this moth, which he named "*Phalaena Neogama*," belongs to a separate species from that described by Linnaeus: "This has been thought not specifically distinct from Linnaeus's *Ph. paranympha*, figured in Roesel, *Vol.* 4 *t.* 18, but we have ventured to presume that it is so from its greater size, the wings being more evidently dentated, with a row of black dots near their margin, the outermost black stripe of the under wings being uninterrupted, and the innermost not recurved back to the base of the abdominal margin."[51] Smith's method involved checking Abbot's drawings against specimens in collections such as Francillon's, William Hunter's, and William Jones's and examining illustrated books devoted to Lepidoptera such as Cramer's *De uitlandsche Kapellen* (1775–82). He also noted verbal descriptions found in Linnaeus's *Systema Naturae* (1758) and Fabricius's *Systema Entomologiae* (1775).

FIGURE 6.11.

Abbot, "Great Yellow-Underwing Moth, *Phalaena Neogama*," bride (*Catocala neogama*), original drawing, TAB. LXXXVIII, Smith and Abbot, *Insects of Georgia* (1797). Courtesy of John Work Garrett Library, Sheridan Libraries, Johns Hopkins University.

Smith firmly believed in the accuracy of Abbot's drawings, saying of a "Brown and Yellow Skipper," "We certainly cannot but trust to Mr. Abbot for the uppermost figure not being a distinct species," and of a Dingy Skipper, "The figure with white spots is like that in Mr. Jones's drawings, described by Fabricius. The other, we presume, on the authority of Mr. Abbot, to be either a sexual difference or a variety."[52] When Abbot's drawings differ from illustrations in Cramer's and Drury's respective books, Smith does not doubt the accuracy of Abbot's depiction; in fact, as in the case of the "Wild-Honeysuckle Hawk-Moth," he ventures to assert that Abbot's drawing represents a new species, which he names "*Sphinx Azaleae*": "We suspect this to be the *Choerilus* of Cramer, though the outline as well as colouring of his are too different from our's to be certainly taken for the same. We have therefore thought it best to give our insect a name applicable to its food, and quote the above figure [Cramer's] with a mark of doubt," meaning that Smith put a question mark after the Cramer citation.[53] Smith's trust in Abbot's drawings to convey the important characteristics necessary for proper identification emboldened him to challenge established nomenclature devised by Linnaeus, the founding father of systematics; his star pupil, entomologist and taxonomist Fabricius; and other published authorities, including Cramer, Rösel, and Drury.

Smith's respectful acknowledgment of Abbot's natural history knowledge and his appreciation of the skill with which Abbot depicted the insects under investigation were not the norm when it came to author/illustrator dynamics and the production of illustrated natural histories. As historians of science Lorraine Daston and Peter Galison have noted, "Battles of wills, eyes, and status" were fought between naturalists and the artists they employed.[54] A brief glimpse at a collaboration between Dru Drury and Moses Harris, an extraordinarily talented artist, a practical naturalist, and author of *The Aurelian*, provides a measure by which we can evaluate the manner in which Smith handled Abbot's visual and verbal contributions to their volumes. For Drury's *Exotic Insects*, Harris made drawings of Drury's specimens, transferred these images onto copper plates, and colored the printed illustrations for many copies of this publication, for which Drury charged his clients an extra fee. Drury provided the text, describing each specimen using Linnaean characteristics and what he knew about the specimen's identity, referring to published authors such as "Clerck, Merian, Roesel, Petiver, &c."[55] Because Drury believed that the value of his book resided in the illustrations' "just and accurate figures," he thought he had to superintend the whole process to guarantee the integrity of his project.[56] He took his role as enforcer of correctness in the coloring of plates very seriously, saying to his readers that if they found any discrepancy between the colors described in the text and the colors in the illustrations to "impute it to its proper cause, the painter."[57]

FIGURE 6.12.

Abbot, "Wild-Honeysuckle Hawk-Moth, *Sphinx Azalea*," azalea sphinx (*Darapsa choerilus*), original drawing, TAB. XXVII, Smith and Abbot, *Insects of Georgia* (1797). Courtesy of John Work Garrett Library, Sheridan Libraries, Johns Hopkins University.

Complaining about "such a multitude of figures which I had to correct," Drury badgered Harris (and, no doubt, his other colorists as well) whenever he felt an illustration was improperly colored.[58] Drury wrote Harris: "I cannot depend on you. . . . Your neglect will prove very injurious to me."[59] Drury, unlike Smith, did not trust his illustrator to represent his specimens correctly. Although Drury was frustrated with Harris's pace and was annoyed at times by what he thought were errors in coloring, this collaboration was ultimately successful. *Exotic Insects*, an endeavor that earned Drury the respect of his peers and the admiration of his fellow entomologists, was considered to be "of very high merit," and as Sir William Jardine wrote in the 1840s, "It is still one of the most beautiful [books] we possess on the subject, and at the time when it appeared" was "almost unrivalled for the elegance and accuracy of its pictorial delineations."[60] Although it might not be fair to compare Drury's conduct with Smith's, since the circumstances surrounding the acquisition of the drawings for their books were quite different, it is worth remarking on Smith's much more respectful treatment of Abbot and his conviction that his illustrations possessed scientific integrity.

Despite Smith's deference toward Abbot, it is clear that he did not hold all practical naturalists in the same high regard. Every so often he lets slip invective against those naturalists less invested in the careful consideration of taxonomic issues: "Too many practical naturalists, and all compiling ones, place species as if no idea of natural affinities ever came across their minds."[61] In moments like this and in his passionate defense of what may seem to be his most arcane taxonomic arguments, Smith's disdain for less theoretical approaches to the study of botany and entomology erupts and threatens to undo the regard he expresses for Abbot. Despite Smith's deference to Abbot's knowledge, it is clear that Smith regarded Abbot as a practical naturalist, as someone who had acquired natural history knowledge through experience but was not knowledgeable or "learned" enough to take up the task of naming new species. Implicit in this distinction between the learned and the practical naturalist was a hierarchy that placed abstract thinking required by classification above empirical knowledge acquired through real-world encounters with living creatures. As the text of *Insects of Georgia* toggles back and forth between Abbot's observations and Smith's systematics, these two voices set up a potential contest for scientific authority between field and study, local and global, situated and abstracted knowledge.

Tensions Within the Entomological Community

Lurking in Smith's dismissal of naturalists who care little about "the philosophy of *things*" is a hierarchy that valued a knowledge of systematics over knowledge gathered in the field. Smith was not alone in this privileging of taxonomy over the field-based activities of practical naturalists, a position that was encouraged by the Linnean Society. This implicit hierarchy was a source of frustration and tension for a segment of the emerging entomological community in Britain, in particular those natural history collectors who did not know Latin. Practical naturalists not privileged with a classical education experienced their lack of training in Latin as a barrier to their full acceptance within London's elite naturalist circles and their scientific societies. Gentlemen, a category that also included physicians and clergymen, received as youngsters an education built on the study of Latin. Naturalists with this kind of education were comfortable with the use of Latin not only to name but also to describe species. Not until the 1770s was Linnaeus's *Systema Naturae* translated into English, and much of the description of characteristics in the emerging field of entomology continued to be written in Latin. For instance, Fabricius, Linnaeus's Danish student, published in Latin, but so did English entomologists Thomas Marsham and William Kirby. The well-worn argument for using Latin was based on its so-called universality, which meant that educated Europeans could read it without having to learn Swedish, Italian, English, or other vernacular languages. Latin was universal in that it could circulate without much friction among classically educated men, but in reality, it acted as a gatekeeper, prohibiting, in effect, most women and men of certain classes from participating fully in scientific work. To counter the tendency within certain scientific circles such as the Linnean Society to ignore the work of practical naturalists not blessed with a classical education, these overlooked insect enthusiasts sought to form an association that would enable them to share what knowledge they had acquired about insects with each other and to improve their understanding of this field of inquiry. These were the reasons for forming the Entomological Society, which was founded in 1806 with the help of a few members of the Linnean Society who believed that entomology would benefit from expanding its base to reach a broader range of potential insect lovers.[62]

The rationale for founding the Entomological Society was explained in a series of published letters issued in response to an ugly attack on this society by Edward Donovan, fellow of the Linnean Society and author of several entomological books, including *The Natural History of British Insects* (1792–1813), *An Epitome of the Natural History of the Insects of China* (1798), *An Epitome of the Natural History of the Insects of India* (1800), and *An Epitome of the Natural History of the Insects of New Holland, New Zealand, New Guinea, Otaheite, and other islands in the Indian, Southern, and Pacific Islands* (1805). In an article published in *The Cyclopædia; or, Universal Dictionary of Arts, Sciences, and Literature*, edited by Abraham Rees, Donovan called the members of the Entomological Society "illiterate" as a way to suggest that their lack of facility with Latin invalidated their attempts at identification. Donovan began his attack with a broadside aimed at Moses Harris's *The English Lepidoptera, or, the Aurelian's Pocket Companion* (discussed in chapter 4) for its use of common English names for butterflies and moths. Donovan's offending paragraph reads:

> The Aurelian's Pocket Companion, by Moses Harris, is a kind of alphabetical catalogue of the larger butterflies, sphinges, and moths, collected in England by its author, a tract of no utility to the Naturalist, unless he first condescends to become an adept in the jargon of the Aurelian, or as now termed, the Entomological Society; the insects being throughout designated by the silly names imposed by the few illiterate persons associated under this denomination. In this English Lexicon of Entomology, we have "The Wall"—"The Bank"—"The Turnpike-gate Keeper"—"Ghost"—"Old Woman"—"Dish-clout [cloth]"—and a vast number of others, equally capricious and replete with absurdity.[63]

The Reverend J. Burrell, a member of the Entomological Society, came to its defense:

> "A few illiterate persons," Mr. Donovan, do you call us! the members of the society are as *literate* as yourself—men of liberal education, and endued with talents equal to your own. . . . Though they may not have squandered so many pounds, as you have done, in the pursuit of this science; though they have not published such errors; yet they are men, who are not illiterate; they are men of sense and education—men who have made as good an use of the means of education, with which God has blessed them, and whose natural capacities have fructified as perfectly as Mr. Donovan's.[64]

Burrell points to Donovan's hypocrisy, as he too had used what he has called "silly names," the English common names for moths and butterflies, in his own publications:

> But who would think, that *Mr. Donovan* would stoop from his dignified literary eminence, and embrace a language he *affects* to despise? Who would, after perusing the article in the *Cyclopaedia*, expect to find his "British Insects" sullified by "Ghost"—"Puss"—"Queen of Soanin"—"Miller of Mansfield"—"Maid of Honour"—"Gipsey"—"Old lady" (a little more genteel than "old Woman!")—"Chimney-Sweeper," and the like. Blush, blush, ye pages of Donovan's British Insects, that your author has tarnished your lustre with such vile trash! Their ugliness will not hide the flagrant errors ye contain, nor divert an eye from the ill-designed, ill-coloured figures, that

> in some many cases present themselves to a purchaser. But I am not to pass sentence on your unfortunate author. I mean only to challenge him to establish his own hypothesis, that the Entomological Society are an illiterate set of beings.[65]

In addition to mocking Donovan's use of common names in his multivolume *Natural History of British Insects*, Burrell critiques Donovan's published illustrations as "ill-designed, ill-coloured figures," a comment that Swainson might have seconded, as he thought Donovan's illustrations were somewhat gaudy.

In another published letter, Reverend Burrell explains how the Entomological Society came into being, noting that several of its "principal promoters" were fellows of the Linnean Society: "These are gentlemen, who unite to their studies in Natural History, the personal labour of collecting the insects of England. Their pursuits and habits throw them into acquaintance with other practical collectors, who were as zealous and diligent as themselves, but not blessed with a classical education—some indeed ignorant of the Latin language, and their information confined solely to Berkenhout, Martin, and other English authors."[66] These "other practical collectors" did not read Latin and were therefore dependent on entomological treatises written in English. Unable to read key entomological texts such as Fabricius's *Philosophia Entomologica*, these practical insect collectors had to rely on texts such as John Berkenhout's *Outlines of the Natural History of Great Britain and Ireland* (1769–72), an English-language text that Abbot also seems to have used as well, judging from a manuscript on English insects that Abbot wrote and illustrated in 1772.[67] Upon examining this manuscript in the 1980s, Ronald S. Wilkinson concluded that "nearly all of the Linnaean content of the manuscript was taken from Berkenhout," but he also points out that Abbot would have had help from Drury, as well as access to Drury's library, which consisted of works by Georg Hoefnagel, Thomas Moffett, Johannes Goedaert, Martin Lister, and Maria Sybilla Merian, all of whom Abbot cited in this manuscript.[68] Clearly, Abbot benefited from interacting with more learned collectors, and this kind of interaction is just what the Entomological Society had hoped to foster.

The founders of the Entomological Society thought that the science of entomology would be better served by collaborating with practical naturalists, who were excellent specimen hunters and often lived outside of London in regions where a greater variety of insects resided.[69] "The object of the Society," according to Reverend Burrell, was "to unite men of respectability who may assist each other in the promotion of the science, and disseminate information to numbers, who labour under the want of a liberal education."[70] However, in defending the Entomological Society from Donovan's "attempts to destroy the credit of the society" and "undermine its fabric," Burrell's description of the society's origins depicts its practical collectors, who "solicited the honour of a more intimate connection" with the learned collectors, as being in need of their guidance, effectively positioning the practical collectors as somehow deficient. By describing the society's goal as "to encourage the practical collector, to lead on the student of moderate education to higher attainments," Burrell unwittingly but ultimately reinscribes the hierarchy based on class dynamics that elevates book-based learning over experience in the field.[71]

To be fair to Donovan, his attack on the Entomological Society was launched from a defensive position. He felt slighted by Adrian H. Haworth, president and founder of the Entomological Society, who described Donovan's work in the preface to his *Lepidoptera Britannica* (1803) in a way that Donovan interpreted as a sly critique:

> The British Insects of my friend Donovan is an elegant and valuable publication; yet it is not a general work, intended to include *complete descriptions of every species*; but confessedly, *a selection only*, of those which are most interesting.
>
> A far more elaborate account of *British Entomology* than that of Donovan, is expected from a pen much superior to either his or mine—I mean that of my friend T. Marsham, Esq. who has laboured a great many years for that purpose, and lately published his first volume, containing the *Order Coleoptera*. The completion of this work will be highly pleasing to all admirers of the science.[72]

Haworth's rapturous praise of the "extreme intricacy" of the "subsequent parts" of Marsham's book seems to have irked Donovan by implying that his *Natural History of British Insects* was not as rigorous or scientific as Marsham's work.[73] Donovan's prickly response and underhanded attack on the Entomological Society can be interpreted as a rearguard defense of an older way of doing natural history with its focus on collecting the rare and the beautiful, categories that connoisseurs valued whether the collected objects were natural history specimens, Italian paintings, Chinese porcelain, or hothouse flowering plants. Donovan had spent his life collecting and drawing rare and beautiful natural history specimens, and he took offense when he recognized in Haworth's description of his and Marsham's work that another approach to collecting was on the rise, one that was based on conducting inventories of insects within a region. All insects mattered in this enterprise, not just the rare and beautiful. In praising Marsham's very technical volume (which is in Latin), Haworth cleared the path for a new way to write books about entomology. However, he saw that a middle ground lay between Donovan's connoisseurial display of the rare and beautiful and the technical treatises in Latin on insect taxonomy such as Marsham's and

Fabricius's. Smith and Abbot's *Insects of Georgia* bridges these two worlds by depicting beautiful and rare Lepidoptera combined with an exacting attention to taxonomy.

These tensions within the entomological community concerning methods for inquiry into insect life were exacerbated by differences in social class, educational background, and professional status. Returning to Burrell's defense of the Entomological Society, we glimpse a problematic term: "amateur."[74] Burrell lumps amateurs with practical collectors: "Every person who is a practical collector, and an amateur of the science may be admitted into this Society, by ballot under regulations bearing great similarity to those of the Linnaean. . . . [A] classical acquaintance with the languages, as I have observed above is unnecessary."[75] Are practical collectors and amateurs two distinct groups, or is there one group that includes members who are not learned and therefore not professional? What did it mean to be a professional naturalist in the early decades of the nineteenth century? The fact that the word "scientist" did not come into the English language until the 1830s suggests how fluid and unstable these categories were, especially when the majority of naturalists were amateurs in the strictest sense of the term, meaning they were not paid for their expertise unless they earned money from publishing their writing on the object of their studies. Yet these "amateurs" were indeed experts who had spent decades studying insects and spiders and writing and publishing about them. An example of the kind of work produced by expert "amateurs" is William Kirby and William Spence's four-volume work, *Introduction to Entomology* (1816–26), considered a landmark in modern entomology. Neither Kirby nor Spence was a professional naturalist like Swainson: Kirby, the rector of Barham, was urged by Spence, a gentleman of independent means, to write an introduction to this new field. Today we tend to associate the term "amateur" with the category of dilettante, someone whose work lacks the mastery we expect from a professional. Clearly, Kirby and Spence's multivolume publication, a founding text of modern entomology, was not the work of dilettantes.

To explore what these terms, "amateur" and "professional," might have meant in this era, I offer another war of words over whose knowledge counts as a way to gauge the valence of these terms in the early decades of the nineteenth century.[76] Forgive this detour into a fight over whether turkey vultures have a good sense of smell that ended in a quarrel about how to preserve bird skins, but it reveals much about the meaning the word "amateur" held within British natural history circles of that time. In "An Ornithological Letter to William Swainson, Esq. F.R.S." (1837), published by Charles Waterton, Esq., of Walton Hall in Yorkshire, Waterton writes: "Sir, I have a crow to pick with you."[77] Apparently, Swainson and Waterton had previously engaged in friendly exchanges over collecting bird specimens, but Swainson had made the mistake of publicly criticizing Waterton's recipe for a mixture designed to preserve bird skins and in the process called Waterton an amateur as a way to dismiss whatever knowledge Waterton proposed to possess.[78] Waterton then defended the category of amateur as someone who did not trade on their expertise and instead shared it with the world for free as he had done by publishing "sixty papers of original observations" in John Claudius Loudon's *Magazine of Natural History and Journal of Zoology, Botany, Mineralogy, Geology, and Meteorology* (1828–40). Waterton zeroes in on the role of money in scientific endeavors, implying that money could cloud one's claims of disinterested curiosity: "If by styling me 'an Amateur', you wish your readers to understand, that I present my zoological information gratuitously to the public;—in contradistinction to yourself, who brings your own to the market;—then indeed, the appellation might be received as a compliment; but, unfortunately, other passages in your work, compel me to view it in no other light than that of a decided sneer."[79] In comparing the "Amateur Naturalist" with the "trading Naturalist," Waterton asserts the priority of knowledge acquired in the field over knowledge acquired by study in museums and libraries and among collections. Flipping the usual hierarchy, Waterton bestows honors on the field naturalist, who in his calculations are amateurs, those who in "the quest for zoological knowledge" go out into the field and "range the dreary swamp;—pursue the wild beasts over hill and dale;—repair to alluvial mud-flats; follow the windings of creeks, and of the sea-shore," and in the process earn "the appellation of 'Amateur' from the pen of supercilious theorists;—an honor, not to be sneezed at, in these our latter times."[80] Waterton's dismissal of Swainson as a "supercilious" theorist is also an implicit rejection of Smith's position that privileged "the philosophy of *things*," the underlying principles upon which systems are built. Swainson had meant to insult Waterton by labeling him an amateur, while Waterton's response turned the term "theorist" into an insult and bestowed honor on the amateur.

Against this background of swirling grievances and thwarted aspirations over whose knowledge had value, we can see that Abbot is somewhat of an anomaly. He does not quite fit into the categories that Smith, Haworth, Donovan, Burrell, Swainson, and Waterton batted about as they defended their positions. Waterton's conflation of amateur and practical naturalist confused matters even more. Abbot was a practical naturalist who, though he did range over dreary swamps and winding creeks, was not an amateur. He earned money by selling specimens and his drawings and was most definitely a professional illustrator, but naturalists were reluctant to acknowledge that artists could possess enough understanding of Linnaean systematics to produce scientifically useful drawings.[81] Fortunately, Smith did seem to appreciate Abbot's careful rendering of insect bodies, which conveyed the necessary details about lepidopteran characteristics required to identify and classify those insects he drew, and Smith used these drawings to identify and name new species.[82] Smith's deference to Abbot's knowledge is remarkable in this era (that a taxonomist and systematist theorist would appreciate Abbot's natural historical knowledge is rare), and their successful collaboration dealt gracefully with the different demands of taxonomy and natural history with its interest in living creatures. Smith's acknowledgment of Abbot's skill in conveying the characteristics important to the classificatory process and his recognition of Abbot's knowledge of insect life did not, however, protect Abbot and the multiple contributions he made to the study of American natural history from being nearly forgotten by the twentieth century.

The Erosion of Abbot's Reputation

In the first half of the nineteenth century, British, French, Swiss, German, and American naturalists who studied entomology knew of Abbot's contributions to entomology as a naturalist, specimen collector, and artist. In 1818 Kirby bought many insects at Francillon's auction; among them were undescribed specimens in the family Fulgoridae (planthoppers) that Abbot had collected.[83] "I was particularly struck by a set collected in Georgia, and admirably displayed by that expert collector, accurate delineator and painter, and most intelligent observer of natural objects." So wrote Kirby in an article published in the *Transactions of the Linnean Society* in which he described (in Latin) the characteristics of two new genera based on specimens Abbot had collected.[84] As previously noted, Swainson, who had had hopes of collaborating with Abbot on a sequel to *Insects of Georgia*, was generous in his praise of Abbot's abilities as a collector and butterfly breeder. Swainson also appreciated the scientific value and aesthetic quality of Abbot's drawings: "Abbot, however, was not a mere collector. Every moment of time he could possibly devote from his field researches, was employed in making finished drawings of the larva, pupa, and perfect insect of every lepidopterous species, as well as of the plant upon which it fed. These drawings are so beautifully chaste and wonderfully correct, that they were coveted by every one."[85] In the 1830s Harvard's Thaddeus Harris corresponded with Abbot, exchanged specimens with him, and collected his drawings, while in 1820s Paris, Baron Walckenaer, the famous arachnologist, finally acquired Abbot's spider drawings after twenty years of trying to persuade Francillon to sell them to him, drawings that were key to his definitive *Histoire naturelle des insectes: Aptères* (1837–47). James Duncan, author of an 1841 volume on exotic moths for Jardine's series *The Naturalist's Library*, noted Abbot's role in "transmitting many fine specimens of insects to private individuals in Britain, as well as to public museums both here and on the continent," while praising *Insects of Georgia* as a very successful joint endeavor: "Of this magnificent publication it is not easy to speak in terms of too high commendation. The drawings were made by Abbot, who was a first-rate entomological draftsman; Sir J. E. Smith superintended the arrangement, and John Harris [Moses Harris's son] engraved the plates: such co-operation could scarcely fail to produce valuable results."[86]

But by the 1880s most of those who had had personal knowledge of Abbot and his work had died, and as a result, recognition of Abbot's contributions to entomology declined to the point that Samuel Scudder in an 1888 article in the *Canadian Entomologist* thought it was necessary to remind his fellow lepidopterists who Abbot was. Scudder expresses admiration for Abbot as a naturalist who "worked to such good purpose and accomplished so much, under circumstances that would now seem very forbidding."[87] Scudder devoted much thought and time to tracking down Abbot's drawings as they moved in and out of French, British, and American collections. However, Scudder's article begins by setting up the distinction between natural history and systematics: "It has been a fortunate thing for the study of butterflies in this country that the earlier students were those who devoted themselves very largely to the natural history of these insects rather than to their systematic or descriptive study."[88] The article proceeds to locate Abbot firmly on the side of natural history and in the process consigns Abbot to a position his readers would have understood as deficient and not quite worthy of the title of scientist.

A major factor in contributing to marginalizing Abbot and forgetting what he had achieved was the fact that the title page to *Insects of Georgia* does not bestow authorship on Abbot. Smith's name appears in the space that is reserved for the author, while Abbot's name is included in the subtitle: "Collected from the observations of Mr. John Abbot, many years resident in that country, by James Edward Smith, M.D. F.R.S." Although Smith does refer to himself in the preface as the "editor" of this book, which is more accurate in that it acknowledges the collaborative nature of the text, Smith's contribution, in effect, eclipsed Abbot's. By naming the new species of Lepidoptera that Abbot had raised and drawn, Smith became the taxon author of the unnamed species in these two volumes. The International Code of Zoological Nomenclature regards this act of bestowing a scientific name on an undescribed species as a form of authorship if published.[89] *Insects of Georgia* contains thirty-five type species, meaning these are the first published names and descriptions of these species. Consequently, whenever one of these thirty-five species is referred to in the scientific literature, a citation needs to follow its binomial name, and this citation follows the author-date system in parentheses.

How, then, is *Insects of Georgia* cited in the scientific literature? Does Abbot's name appear in the citation as one of the authors of this book? In 1819 "Smith-Abbot," according to the French *Encyclopédie méthodique par ordre des matières* (Methodical encyclopedia by order of subject matter) (1782–1832), was the correct way to cite *Insects of Georgia*, and in the 1840s Duncan referred to *Insects of Georgia* as authored "by John Abbot and Sir J. Edward Smith."[90] But over the course of the next one hundred years, the use of Abbot's name declined in references to *Insects of Georgia*. In 1958 Cyril dos Passos complained in the *Lepidopterists' News* that "the authorship of the scientific names" in *Insects of Georgia* had been "ascribed variously to ABBOT, ABBOT and SMITH, SMITH and ABBOT, and more recently to SMITH."[91] Because dos Passos was in the process of publishing a checklist of a region's North American Lepidoptera, he needed to know how to refer to the species described and identified in *Insects of Georgia*. He ended up choosing "Smith, 1797." "But," he writes, "in a synonymy could properly be followed by '*in* Smith and Abbot, 1797,'" which signals dos Passos's recognition of Smith's and Abbot's different responsibilities as authors.[92]

Dos Passos's distinction between authoring taxa and authoring a book is echoed on the marvelous website "Butterflies and Moths of the World," built in the 2010s by London's Natural History Museum. The website's list of thirty-five type species depicted in *Insects of Georgia* is credited to Smith. This is as it should be if the goal is to acknowledge who published the first scientific name and description of a species. But where is Abbot in all of this? The website manages to also make room for Abbot, capturing the complexity of Abbot's and Smith's collaboration. For instance, the binomial name *Sphinx juglandis* is followed by "(Smith, 1797)," indicating that Smith was the first to publish its scientific name, but right after this is a reference to *Insects of Georgia*: "In: Smith & Abbot, *Nat. Hist. lepid. Ins. Georgia* 1: 57, pl. 29."[93] This simple but important acknowledgment of Abbot's role as an author of *Insects of Georgia* does not undermine Smith's status as the systematic author; instead, it brings Abbot's contributions as a natural history author back into view, mitigating to some small degree his absence in the scientific literature and the history of American natural history.

In general, however, Abbot was too often characterized as a mere source of information for what is regarded as Smith's *Insects of Georgia*. Smith, who did not know all that much about insects, came to be regarded as an authority on butterflies and moths, an authority he would never have enjoyed if it had not been for his collaboration with Abbot. What Smith did have, aside from his wealth, education, and social connections, was Linnaeus's insect collection. He purchased it from Linnaeus's widow along with Linnaeus's other natural history collections and his library; these purchases Smith gave to the Linnean Society, an organization he founded in 1788. Smith also had access to Joseph Banks's library and specimens, as well as Kew Gardens' herbarium. This access, combined with the natural history expertise located within his circle of Linnean fellows (Macleay, Francillon, Kirby, Jones, and Haworth), put Smith in a position to do the work of identification, naming, and classification. Abbot, even if he had been so inclined, would not have been able to participate in this process, as he worked alone and far from the centers of natural history expertise with their librarians, curators, and massive specimen collections. Without the recognition that scientific authorship bestows, Abbot's name became disassociated from the book that would not have existed without his drawings and notes. Further eroding Abbot's reputation as one of the most knowledgeable and talented natural history illustrators of his era was the publication of a letter by Augustus G. Oemler in an article by Robert Percy Dow published in the *Journal of the New York Entomological Society* in 1914. As we shall see, Oemler's letter combined with Dow's interpretation of it had a profoundly negative impact on Abbot's reputation as a naturalist in the twentieth century.

7

Abbot's Legacy

IN 1914 ROBERT PERCY DOW published an underhanded attack on Abbot's credentials as an accomplished naturalist and natural history illustrator. He portrayed Abbot as a mere dealer in specimens, claiming that the artist was generally both uneducated and ignorant of Linnaean nomenclature and systems of classification and, for these reasons, was unworthy of his association with Smith. The evidence Dow marshaled to make this argument was drawn from a letter written in 1834 by Abbot's friend Augustus G. Oemler, a letter that was sent to the Harvard librarian Thaddeus W. Harris, a naturalist with whom Abbot had been corresponding. Apparently, Harris had inquired about Abbot's personal history, and Oemler complied with his request, not realizing that eighty years later, Dow would use this letter as a weapon to undermine what remained of Abbot's reputation as a knowledgeable and gifted natural history artist.

This chapter confronts the marginalization of Abbot's achievements as a naturalist in the twentieth century by revisiting Oemler's letter and analyzing Dow's interpretation of it. To counter Dow's depiction of Abbot as lacking scientifically valuable knowledge of North American arthropods, I then turn to the example of how Baron Walckenaer relied on Abbot's spider drawings and notes to realize his ambitious undertaking, the multivolume book *Histoire naturelle des insectes: Aptères* (1837–47), which became a foundational text in the study of arachnids. Using the example of Walckenaer's dependence on Abbot's deep knowledge of spiders, this chapter attempts to undo the damage wreaked by Dow and those who repeated his unwarranted accusations and to insist on the importance of Abbot's contributions to the scientific study of the natural world.[1]

"The Simplicity of a Boy"

In the 1830s Harris had asked Oemler if he could share some personal anecdotes about Abbot, and Oemler responded with what has become a much-cited letter. Oemler told Harris that he had first met Abbot in 1805 and that he had last seen

Abbot in the mid-1820s. He reported that he had found Abbot well but quite deaf: "I had to use the Slate to converse with him. But notwithstanding this misfortune, he was cheerful, and his constitutional smile never left his countenance. He is now, I am told, very corpulant, but still exercises his pursuit of hunting birds and drawing—but engages boys to run after Butterflies." Oemler reminisced about how Abbot "never knew any thing of Linneus' Classification till I demonstrated it to him and created his astonishment." With this statement, Oemler takes credit for teaching Abbot how to draw plants with the proper number of pistils and stamens to conform to the Linnaean system of botanical classification. He made much of Abbot's ignorance of Linnaean systematics: "You will be astonished when you hear that a man, so long amusing himself with Natural History, should never have been inclined to pursue it scientifically, he, although now 83 years of age, is still in the simplicity of a boy." Despite these comments, Oemler tells Harris that he is determined that "Abbot should be known to the world. . . . [T]he Laurals, which are assumed by Smith and Leconte should be placed on the proper brow," and he encourages Harris to see that "justice be done him."[2]

Oemler's letter is riven by contradictions. On the one hand, Oemler believed that Abbot deserved to be as well known and as honored as Smith, the president of the Linnean Society, and LeConte, who coauthored with Boisduval the influential book on North American Lepidoptera filled with illustrations by Abbot. On the other hand, he portrayed Abbot as a simpleton, ignorant of Linnaean systematics and amazed by the revelation that he needed to paint the number of pistils and stamens in a flower accurately. Abbot, of course, was not a simpleton, nor was he ignorant of Linnaean systematics.

Harris died in 1855, and Oemler's letter lay hidden for decades in the attic of Harris's Cambridge, Massachusetts, house, tied up in a bundle of two hundred letters that had been sent to him by various correspondents. Oemler's letter did not see the light of day until 1913, when the Harris family home was searched by eager entomologists looking for documents that may not have been conveyed to the Boston Society of Natural History, the repository that held most of Harris's papers.[3] A year later Dow published Oemler's letter in an entomology journal article, "John Abbot, of Georgia," and proceeded to assess whether Abbot deserved those laurels that Oemler had wished to place on Abbot's brow. Dow repeated and amplified Oemler's claim that Abbot did not know about Linnaean systematics until Oemler taught him: "He never heard of the Linnean system until after 1805."[4] Dow depicted Abbot as an "untutored" observer, minimizing Abbot's scientific contribution to *Insects of Georgia* and the study of North American Lepidoptera. Twentieth-century scholars who relied on Dow's article concluded that Abbot "never heard of Linnaeus, nor of any form of classification, until he was well over 50" and that he "lacked instruction in scientific thought."[5]

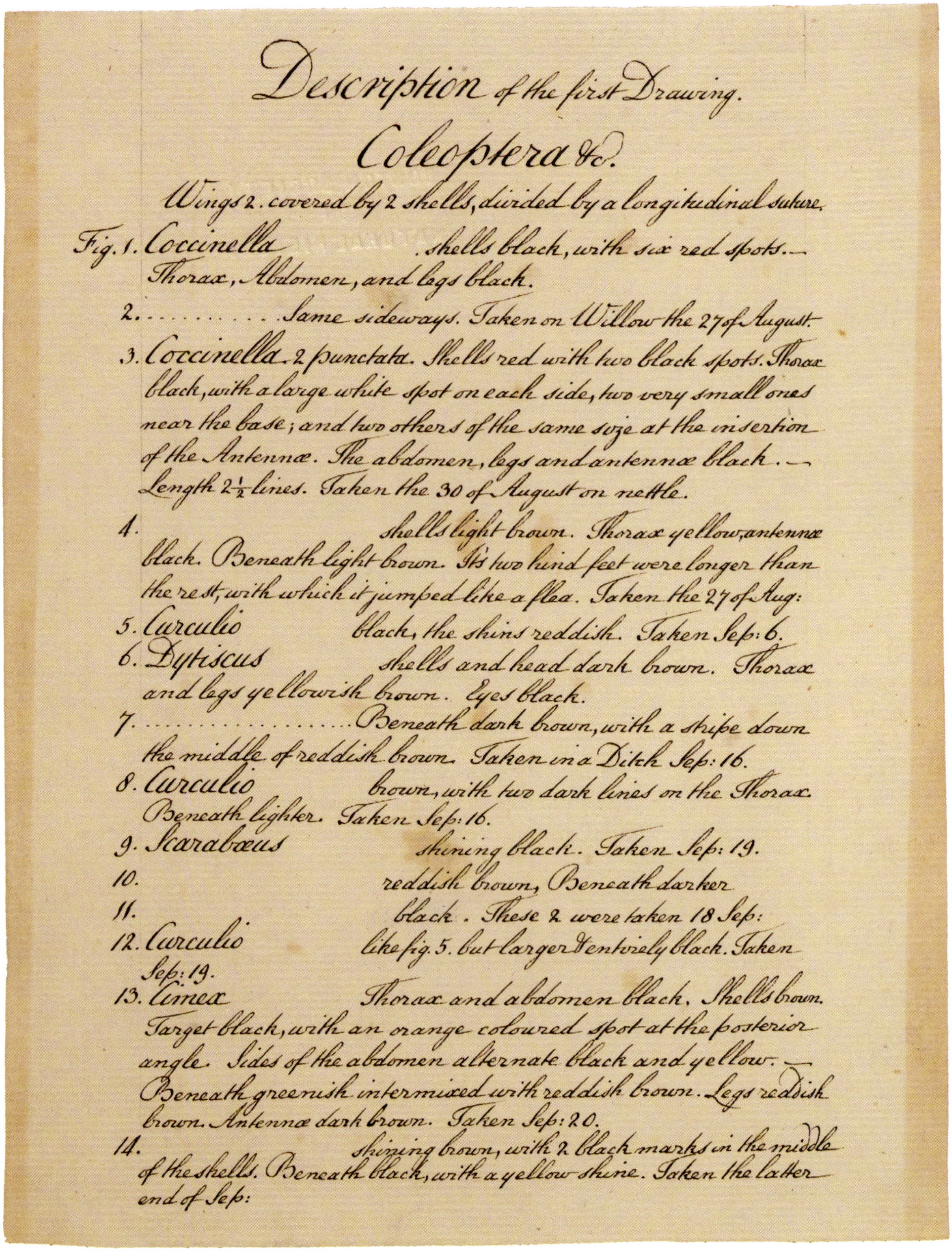

Description of the first Drawing.

Coleoptera &c.

Wings 2. covered by 2 shells, divided by a longitudinal suture.

Fig. 1. Coccinella .shells black, with six red spots.— Thorax, Abdomen, and legs black.

2. .... Same sideways. Taken on Willow the 27 of August

3. Coccinella 2 punctata. Shells red with two black spots. Thorax black, with a large white spot on each side, two very small ones near the base; and two others of the same size at the insertion of the Antennæ. The abdomen, legs and antennæ black.— Length 2½ lines. Taken the 30 of August on nettle.

4. shells light brown. Thorax yellow, antennæ black. Beneath light brown. It's two hind feet were longer than the rest, with which it jumped like a flea. Taken the 27 of Aug:

5. Curculio black, the shins reddish. Taken Sep: 6.

6. Dytiscus shells and head dark brown. Thorax and legs yellowish brown. Eyes black.

7. .... Beneath dark brown, with a stripe down the middle of reddish brown. Taken in a Ditch Sep: 16.

8. Curculio brown, with two dark lines on the Thorax. Beneath lighter. Taken Sep: 16.

9. Scarabæus shining black. Taken Sep: 19.

10. reddish brown, Beneath darker

11. black. These 2 were taken 18 Sep:

12. Curculio like fig. 5. but larger & entirely black. Taken Sep: 19.

13. Cimex Thorax and abdomen black. Shells brown. Target black, with an orange coloured spot at the posterior angle. Sides of the abdomen alternate black and yellow.— Beneath greenish intermixed with reddish brown. Legs reddish brown. Antennæ dark brown. Taken Sep: 20.

14. shining brown, with 2 black marks in the middle of the shells. Beneath black, with a yellow shine. Taken the latter end of Sep:

FIGURE 7.1.

Abbot's description of beetles with Linnaean nomenclature. John Abbot, "A Natural History of Insects" (1772), MS Typ 426.1, Houghton Library, Harvard University.

Neither of these assertions is true: Abbot had labeled his drawings using Linnaean nomenclature as early in his career as the late 1760s and early 1770s, when he was in his late teens and early twenties.[6] (See figure 7.1.)

A closer look at Oemler's letter and Dow's article can help explain in part how Abbot disappeared from twentieth-century accounts of the history of American natural history and specifically from the scientific literature that cites *Insects of Georgia*. Oemler may have believed that he was the one to have explained Linnaean systematics to Abbot, but we know that Abbot spent years under Drury's tutelage, during which he most certainly learned what characteristics naturalists needed to see in order to identify and classify insects according to Linnaean systematics. Drury, an "avid disciple of Linnaeus," explained in the preface to his *Exotic Insects* that Linnaeus was the principal author whom "I have quoted among the descriptive writers: his great judgment in this study, the plain method he has laid down for the classing of insects, together with the excellency of his generical characters, are what must endear him to every professor of this study."[7] Abbot was certainly privy to these sentiments, as his tenure in Drury's circle coincided with the preparation and publication of Drury's magnum opus. We also know that in 1770 he purchased a copy of Drury's first volume that had been colored by Moses Harris, the book's illustrator and engraver.[8] As for the scientific accuracy of Abbot's drawings of plants, would Smith, a botanist, taxonomist, and promotor of Linnaean systematics, have described the drawings for *Insects of Georgia* that featured lepidopterans and their host plants as "the most *perfect* thing I ever saw" if he thought they were deficient in their adherence to Linnaean systematics?[9] Abbot's habitual modesty about his skills and knowledge may have misled Oemler, and Abbot may have let Oemler believe that he had instructed him in the proper way to draw plants. After all, Abbot had characterized himself in 1819 as "no Botanist, only an admirer of Nature[']s Beauties, to meet with a new growing flower or plant much pleases me."[10]

Oemler's portrait of Abbot as having the "simplicity of a School boy" might not necessarily have been intentionally demeaning; instead, it might merely have been his way of describing Abbot's delight in finding and drawing insects and the way in which Abbot expressed the wonder he experienced in his encounters with the natural world. There may, indeed, be something childlike about a grown man running through tall grasses waving a butterfly net or plopping down on the ground in a forest to take a closer look at a beetle scurrying under a decaying log. Oemler's letter, it can be argued, might not have been written in bad faith, but in Dow's hands it became a source of misinformation that circulated widely, with damaging consequences for Abbot's legacy.

Although Oemler did say that Abbot "never knew any thing of Linneus' Classification" and that he did not pursue natural history "scientifically," it can be argued that Dow had misread Oemler's letter, which was most likely refer-

geometra

ring to what Oemler thought was Abbot's lack of familiarity with systematic botany. As Ronald S. Wilkinson argues, "The only explanation to fit the facts is that Oemler's remarks about unscientific pursuit and simplicity were principally directed to the area of natural history with which *he* was almost entirely concerned; in the unfortunate and inaccurate generalization, as well as the specific example of the stamens, he was criticizing Abbot's botanical knowledge." Although there might be a grain of truth in Oemler's assertion (but only as it applies to botany), Wilkinson contends that Abbot "pursued the study of insects and birds in a manner which, in the sense of his time, can only be characterized as scientific."[11] Wilkinson laments that two fine scholars, taken in by Dow's mischaracterization of Abbot's scientific training, spread this misinformation in the 1970s and 1980s:

FIGURE 7.2. Abbot, "Brown and White-Margined Moth (*Phalaena Calycanthata*)," double-banded zale (*Zale calycanthata*), original watercolor drawing, TAB. CIV, Smith and Abbot, *Insects of Georgia* (1797). Courtesy of the John Work Garrett Library, Sheridan Libraries, Johns Hopkins University.

> The admirable Charles Cowan has suggested that Abbot "made friends with several prominent naturalists in London (though he never heard of Linnaeus, nor of any form of classification, until he was well over 50)." Vivian Rogers concluded that Abbot "lacked instruction in scientific thought. He grew up unaware of Carolus Linnaeus's system of consistent binomial nomenclature for species of plants and animals which was first advanced in 1753. It was not until a Savannah friend, Dr. Augustus G. Oemler, demonstrated it to him some time after 1805 that he learned of it." . . . Both writers appear to have depended on a 1914 paper by Robert P. Dow.[12]

Wilkinson suggests that if Cowan and Rogers had examined Abbot's "eighteenth-century drawings and notes," they would have noticed Abbot's "awareness of Linnaeus' binomial nomenclature when he was very young, and his use of the zoological classification as early as 1772."[13] These drawings and notes referred to by Wilkinson (and discussed in chapter 4) are held in Harvard's Houghton Library and are open to the public; they are now digitally available online.[14]

There seems to me to be more to Dow's contention that Abbot never heard of the Linnaean system of classification until he was in his midfifties than simply misconstruing Oemler's intention. He seems determined to portray Abbot as ignorant and uneducated; "untutored" is his term. Dow asserts: "His education was limited. His grammatical blunders would be unpardonable in any grade above elementary school. His name does not appear in the matriculation lists of any university or great public school in the United Kingdom. . . . He was not a member of the short-lived Entomological Club of London, 1781–3."[15] This last point is meaningless, since Abbot was not in England in the 1780s. Nor does Dow seem to know much about higher education in eighteenth-century England, where the university curriculum did not prepare one to be a scientist. Not until the 1860s did the entomologist John Obadiah Westwood (1805–93), the curator of the Hope

Collection at Oxford University, earn a professorship of zoology. Knowledge about the natural world was acquired not in the academy but through informal and experiential learning that included organizations made up of like-minded enthusiasts where initiates were mentored much the way Abbot was mentored by Drury.[16] Also misinformed is Dow's emphasis on Abbot's poor spelling as a gauge by which to measure his intellectual stature: "The misspellings of Abbot reveal the man."[17] Spelling in the eighteenth century was a fluid affair with little concern for correctness. (For example, Sir Joseph Banks, president of the Royal Society, was a terrible speller.) General-purpose English-language dictionaries were a relatively new thing in the eighteenth century. (Dr. Johnson's dictionary, for instance, was first published in 1755.) Abbot's formal education was in the form of an apprenticeship, working as a law clerk for his father, and his natural history education was shaped by Drury and those in his circle.

With these misinformed allegations concerning Abbot's education, Dow appears to be determined to dismiss Abbot's accomplishments as a naturalist. Unlike Oemler, who had wished to see "justice be done him" and that the laurels "should be placed on the proper brow," Dow seemed more interested in stressing Smith's rightful claim to the laurels that Oemler had thought should have been Abbot's: "Let us be fair about the laurels not on Abbot's brow. It was certainly Smith's privilege to give scientific names to these undescribed species. He bought types, as well as drawings from Francillon. If you or I bought from a New York dealer two dozen butterflies from Africa, raised there by a dealer who sold them for profit, and who was not known to you save by name alone, and you found them to be new species, would you credit his name with yours as the describer?"[18] The answer Dow is fishing for in this passage is "no." Why would you give credit to a dealer for discovering and naming a new species when you did all the work of classifying the unnamed species? Dow goes on to say that Smith's giving credit to Abbot as "the worker in the field is all that the finest sense of honor can demand."[19] With this rhetorical move, Dow turns Abbot into a supplier of specimens, which he was, but he was, of course, much more than that. In another article Dow states that Abbot "was not a coleopterist, nor was he a scientist, even, in the ordinary acceptance of the word." Abbot, Dow contends, "observed closely and drew beautifully a large number of lepidoptera and plants. Of their generic or specific [as in genus and species] determination he knew little."[20]

In contrasting observation with systematics, Dow calls upon the well-worn distinction we have seen at work within nineteenth-century entomological circles, the distinction between theory-driven knowledge and empirical knowledge produced by observation. Natural historical knowledge that was based on observing living animals and plants in their native habitats was often depicted as somehow inferior to the kind of knowledge that was generated by systematics, which in the eighteenth and nineteenth centuries required the study of textual descriptions of collected specimens, something that was done indoors in a museum setting or what was called a "closet," meaning a private space designed for study and reflection where, in this case, a naturalist surrounded by specimens and reference books would do the work of classification.[21] Dow's description of Abbot's skills as an artist and an observer was meant to put him in his place, which Dow believed was very much beneath Smith, who, with his command of Linnaean systematics and access to zoological books and specimens, was an expert (botanical) taxonomist.

The Scientific Value of Drawings

By depicting Abbot as an "untutored" observer, Dow minimizes Abbot's scientific contribution to *Insects of Georgia* and the study of North American Lepidoptera. Of the myths about Abbot, which were mostly generated in the late nineteenth and early twentieth centuries and which continue to circulate today, the most damaging is the idea that because he was not formally trained as a scientist, he did not realize the significance of what he had produced and that his skills as a careful observer were enough to render his drawings of use to the entomologist and the taxonomist. This idea that an uneducated eye could produce such careful renditions of insects and spiders is deeply naive. Dow did not understand that naturalist artists such as Abbot possessed highly trained and selective ways of seeing that allowed these artists to produce "reasoned images."[22]

Under Drury's tutelage, Abbot had learned Linnaean nomenclature and insect morphology, tools that enhanced his ability to convey visually what other naturalists wanted to see. Abbot attended to insects in highly specialized ways of seeing, focusing on an insect's anatomy and looking for those characteristics—wing venation, segments and spurs on legs, palpi and mouth parts, number of segments on antennae, and so on—that were deemed important to identification within Linnaean systematics. Of the Lepidoptera Abbot drew, Thaddeus Harris thought that the species were "easily identified by Abbot's figures, although from Smith's descriptions alone I could not have made out half of them."[23] Abbot's insect drawings are so steeped in Linnaean systematics that current-day entomologists, who still use many of the same characteristics of insect morphology that Linnaeus and Fabricius used to identify insects, can identify all of the butterfly species and the majority of the moth species depicted in his drawings, as well as the genera and many of the species of the beetles he drew. Because Abbot's drawings represent his thinking about how to classify the insects he had caught, killed, and preserved, his art making can thus be understood as an exercise in thought.[24] Not merely illustrative, Abbot's drawings were actively engaged in the complex cognitive functions that structured the taxonomic thinking central to the emerging discipline of entomology.

So painstakingly careful in his rendering of whole (not dissected) insect bodies in life-size detail, Abbot produced drawings that several of the era's leading entomologists were eager to own. As mentioned earlier, Swainson stated that Abbot's drawings were much desired, "so beautifully chaste and wonderfully correct, that they were coveted by every one."[25] Evidence of the high esteem with which entomologists regarded Abbot's drawings can be seen in an anecdote involving the efforts to acquire Abbot's drawings of spiders made by one of the great arachnid taxonomists of the nineteenth century, Baron Walckenaer. In the preface to his *Histoire naturelle des insectes: Aptères* (1837–47), a foundational text in the study of arachnids, Walckenaer tells the story of how difficult it was for him to get his hands on Abbot's drawings of spiders and the accompanying notes, items he needed to help compile his magnum opus. He explained that there were two manuscripts that had proved invaluable to this process, one by Louis Augustin Guillaume Bosc that describes 25 spider species from the Carolinas and the other by Abbot (who is misidentified as Thomas) with 535 drawings of spiders "d'après des individus vivants" (from living individuals). Walckenaer described how he acquired Abbot's drawings and accompanying notes:

> These drawings of Araneids and Phalangids were accompanied by a notebook of forty-two pages written in English entitled *Notes and Observations on the Drawings of the Spiders of Georgia*. These notes and observations contain short descriptions of the species painted by the author. When I learned that Mr. Abbot, who had spent five years collecting the material for this work and executing it, having the intention to sell it, I wished to acquire it, and to achieve this I began in 1802 a correspondence on this subject with Mr. John Francillon of London, who had Abbot's manuscript in his possession and was charged with the sale on behalf of the author. Mr. Abbot lived in Georgia then. This affair could not be concluded because he [Francillon] refused to send the manuscript to Paris. Mr. MacLeay, the celebrated entomologist, became its owner. He was kind enough to put it in my hands, and after having examined it, I acquired it in 1821.[26] (my translation)

Why did Francillon, who brokered many sales of Abbot's drawings, refuse to sell his spider drawings to Walckenaer? Perhaps he was unwilling to send the drawings to Paris for Walckenaer to examine before the French arachnologist would commit to purchasing them. The political instability of France, given the rise of Napoleon and the ensuing warfare that engulfed most of Europe after 1805, may have led Francillon to conclude that it was an unsafe place to send

FIGURE 7.3.
Abbot, "Spider +235," wolf spider (*Gladicosa pulchra*), det. J. K. Wadleigh, watercolor drawing, "Aranea" (ca. 1799), Hargrett Rare Book and Manuscript Library, University of Georgia Libraries.

such valuable manuscript material. Five years after Francillon's death in 1816, Walckenaer was finally able to purchase a set of Abbot's spider drawings from Alexander Macleay.[27]

Francillon's reluctance to sell Abbot's drawings to Walckenaer seems even stranger when it is quite clear that Abbot was eager to find a market for his spider drawings. Writing to Swainson in 1816, Abbot asked him if he knew of anyone who might like to buy them, "the Spiders of Georgia being numerous and many remarkably for size and beauty. I have an Intention of Drawing a set of 5 on a Q[uarto] size Drawing in the manner of Albini [Eleazar Albin] if You or any friends should be desirous to purchase them. I have no doubt we shall agree as to the price, I therefore leave it to You or them to do it, presuming you or they will readily give me what is reasonable for my Trouble."[28] Swainson's reply is somewhat insulting, as he suggests that Abbot might not know how to delineate spiders correctly, casting doubt on his ability to contribute meaningfully to the scientific study of arachnids. Swainson replied to Abbot's query: "I cannot decide on taking the drawings of Spiders you mention, until I see two or three as a specimen. I need not mention to you that the greatest accuracy is necessary in delineating the position of their eyes for by their [that] character they are now divided into distinct genera. If you will send me two or three drawings I can then decide, and if I do not engage with you myself, will use every endeavour to serve you among my friends."[29] Of course, Abbot knew that eye configuration was key to arachnid identification. Above most of the figures of spiders in his drawings, Abbot placed a few dots to convey the position of their eyes to aid in spider identification and classification. Above these dots and the image of the spider Abbot wrote a number that corresponded to numbers on a list that made up his notes. Next to each number in these notes was a description of the pictured spider: how, when, and where he caught it; what it ate; what its web looked like (if it had a web); how rare or common it was; and any interesting local knowledge about these species. The minuteness of his subject matter did not deter him from giving these spiders his full attention as a draftsman. He recorded with precision the patterns on these spiders' abdomens, using paintbrushes consisting of only a few hairs to carry microscopic dots of watercolor paint onto the paper. In doubting Abbot's ability to draw spiders accurately, Swainson underestimated Abbot's skills as an artist and a naturalist.

Abbot's Influence on Walckenaer

Abbot's spider drawings and notes were the product of years of studying spiders in the region between the Ogeechee and Savannah Rivers and are essentially an inventory of the five hundred plus species of spiders living in Southeast Georgia in the late eighteenth century. For Walckenaer, these materials were invaluable for his encyclopedic project of describing and naming every known spider species. Walckenaer's work, *Histoire naturelle des insectes: Aptères*, is filled with quotes from and citations to Abbot's notes. Volume 1 contains eighty-nine references to Abbot's notes, and volume 2 has seventy-eight references. For instance, Walckenaer relied solely on Abbot's drawing and description of "The Ground-Hole Spider," a wolf spider, naming it "Lycosa Tarentule Georgienne" (now *Tigrosa georgicola*).[30] Another wolf spider, which Walckenaer named *Lycosa avida* (now *Schizocosa avida*), was pictured by Abbot as figure 46 on plate 10 of his manuscript and described on page 7 of his notes. Walckenaer included Abbot's thoughts on this wolf spider in his description: "Abbot dit qu'en dessous le ventre et les hanches sont d'un noir de velours, et que les mandibules sont convertes de poils jaunes, variable. Abbot a pris cette Aranëide courant par terre, dans un bois planté de pins, le 27 octobre" (Abbot says that the underside of the abdomen and the coxa and trochanter [hips] are velvety black and that the mandibles are sometimes covered with yellow hair. Abbot took this spider running on the ground in a pine forest on October 27).[31] The information contained in Abbot's drawings and observations of spiders enabled Walckenaer to describe and name hundreds of new species of North American spiders.[32]

In the 1930s two American arachnologists, Ralph V. Chamberlin and Wilton Ivie, sought Abbot's spider notes and drawings to help them with their inventory of spiders in Georgia. Inspired by Walckenaer's use of Abbot's work on spiders, they realized that if they could examine Abbot's drawings, they might be able to identify spiders that Walckenaer had not classified. In a 1944 article they explained that Walckenaer had utilized Abbot's drawings to name 282 new species of the 567 pictured spiders.[33] Working from Abbot's drawings held in London's Natural History Museum (Francillon's volume 14), Chamberlin and Ivie were able to identify another 165 "with tolerable certainty."[34] Between Walckenaer's list of identifiable species and their own, the American arachnologists developed a baseline of spiders inhabiting Southeast Georgia in the late eighteenth century, a baseline against which they would measure the results of their own inventory. Over the course of two summers in the mid-1930s they conducted an inventory of spiders, and then in 1943 they collected more specimens, focusing on "the vicinity of Sylvania and Savannah, Georgia, with a brief excursion into the more mountainous territory between Gainesville and Clayton. Sylvania, in Screven Co., is near the place where Abbot did most of his collecting, and for that reason was chosen as the principal base. On this last expedition 350 recognizable species were taken."[35] Unsurprisingly, they found fewer species than Abbot did. They believed that the lower number was a result of the "absence of autumn collecting," although we might today attribute some of the missing eighty-eight spider species to habitat loss and other changes in the environment. Chamberlin and Ivie wondered why there had been so little research done on this region's spiders: "This has been due to lack of collecting there and not to scarcity of spiders, for undoubtedly it can hold its own in number of species with any region of North America north of 30° latitude."[36]

Chamberlin and Ivie assumed that Walckenaer used the spider volume in Francillon's collection held in London's Natural History Museum. However, they were wrong, not realizing that there was another volume of spider drawings on which Walckenaer had based his work. This volume now resides in the library of the National Museum of Natural History in Paris, and it contains proof that Walckenaer used this volume and not the album in London. Tucked inside the Paris volume are letters from Francillon and Walckenaer's correspondence that support the baron's story about how he had acquired this volume only after Francillon's death. Not realizing that there are two different spider volumes, Chamberlin and Ivie thought Alexander Macleay had delivered to Walckenaer the spider volume from Francillon's collection after the jeweler's death, and then after the baron's death, the volume somehow made its way to London and into the Natural History Museum's collection.

After the baron died in 1852, his library was auctioned off on April 12, 1853. Lot 513, a set of drawings, is described in the auction catalog as a "Collection of 535 different species of spiders from the state of Georgia in the United

FIGURE 7.4.
Abbot, orb weaver or ally (Araneoidea sp.), bark crab spider (Coriarachnini sp.), fishing spider (*Dolomedes* sp.) (*center*), orb weaver or ally (Araneoidea sp.), orb weaver or ally (Araneoidea sp.), det. J. K. Wadleigh, watercolor drawing, "Abbot's Georgian Spiders" (1798), drawing no. 1, MS 274, Muséum national d'Histoire naturelle, Paris.

States of America, drawn and colored with care by Abbot. This precious volume was bought in London for 600 francs by M. Walck. as attested by several letters in French and English, which had been added to the volume."[37] The preface to this auction catalog describes the scientific importance of Walckenaer's library, explaining that his study of spiders required reference to the works of René Antoine Ferchault de Réaumur, Johann Christian Fabricius, Carl Linnaeus, Eleazar Albin's *Natural History of Spiders*, Thomas Martyn's republication of that volume along with Carl Alexander Clerck's *Aranei Svecici* (1757), "and another work of the highest importance, which perhaps should have been cited first (no. 513); Abbot, collection of 535 species of spiders from Georgia in America."[38]

To add to the confusion over these two sets of spider volumes, there is a third set of Abbot spider drawings that differs in size and layout from those in the London and Paris volumes. The London volume has 582 images of arachnids; 567 of these are "Aranea" (spiders), and 15 are "Phalangium" (harvestmen). The format of these spider drawings is quite unlike that of any of the drawings in the other Francillon volumes, the majority of which are made up of single specimens on small sheets of paper. The layout of these spider drawings in volume 14 is similar to that of Abbot's drawings in Avinoff's collection, with their decorative display of several arthropod bodies on one sheet of paper. Likewise, the Paris volume is made up of 535 arachnids pictured on 107 pages, the majority of which is made up of spiders, and each page has a geometrical arrangement of five arachnids pictured on it.

The third spider album, held at the University of Georgia's Hargrett Library, contains 539 drawings of arachnids and had once been the property of Manchester's Chetham Library, which had been Francillon's client. Unlike the Paris and London albums with their geometrical arrangement of several spiders on each page, each spider in the Hargrett album is on its own small sheet of paper, and three to four of these sheets of paper are pasted onto a folio sheet of laid paper with a J. Whatman 1794 watermark. Accompanying these drawings are Abbot's observations, which are almost word for word the same as those passages Walckenaer quoted in his book and nearly the same as those Francillon transcribed into his spider volume.[39] There are also 126 spider drawings at the American Philosophical Society that make up a portion of the eight volumes that had been compiled by the LeContes.[40]

It has been over eighty years since Abbot's spider drawings inspired Chamberlin and Ivie to conduct inventories of Georgia's spider populations. Should some enterprising arachnologists take up the mantle and conduct another inventory of Georgia's spiders, Abbot's spider drawings and notes could be used to create a baseline against which to measure current loss and the environmental pressures on biodiversity.

Abbot ad vivum delin. 1798.

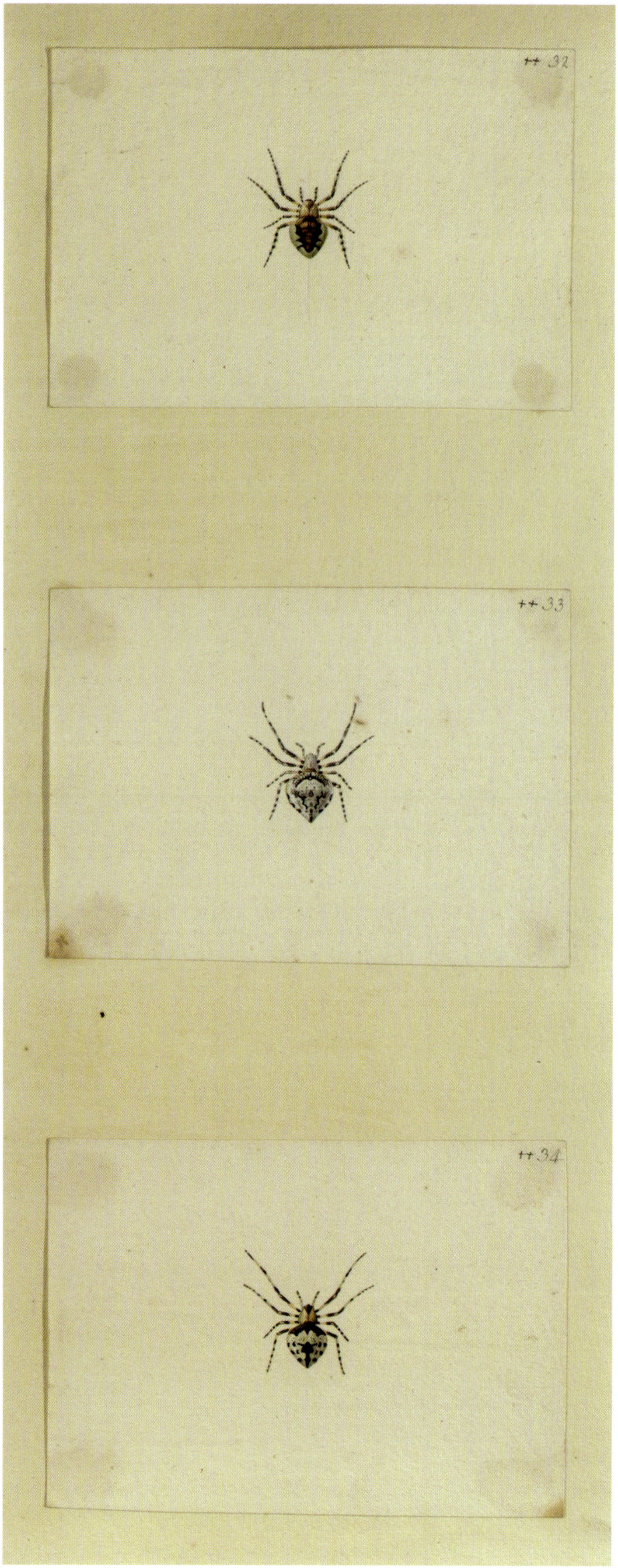

FIGURE 7.5.
Abbot, "Spiders ++32, ++33, ++34," humpbacked orb weaver (*Eustala anastera*), det. J. K. Wadleigh, watercolor drawing, "Aranea" (ca. 1799), Hargrett Rare Book and Manuscript Library, University of Georgia Libraries.

FIGURE 7.6.
Abbot, "Spider ++27," star-bellied orb weaver (*Acanthepeira stellata*), det. J. K. Wadleigh, watercolor drawing, "Aranea" (ca. 1799), Hargrett Rare Book and Manuscript Library, University of Georgia Libraries.

FIGURE 7.7.
Abbot, "Spider ++219," female white-banded crab spider (*Misumenoides formosipes*), det. J. K. Wadleigh, watercolor drawing, "Aranea" (ca. 1799), Hargrett Rare Book and Manuscript Library, University of Georgia Libraries.

"Father of American Arachneology"

Recognition of the significance of Abbot's contribution to the study of North American arachnids has been slight despite Walckenaer's forthright acknowledgment of his indebtedness to Abbot. His contribution to the study of spiders has been generally forgotten, even among arachnologists, primarily because Walckenaer's work did not receive the recognition it deserved in the United States in the nineteenth century. Much of his work on North American arachnids was eclipsed by Nicholas Marcellus Hentz (1797–1856), a French immigrant to the United States who published an article in the *American Journal of Science and Arts* in 1832 that listed 125 new North American arachnid species.[41] With this publication, Hentz beat Walckenaer in the race for naming rights of these 125 species. Even though Walckenaer published volumes 1 and 2 of *Histoire naturelle des insectes: Aptères* in 1837, both of them included more species than Hentz's 1832 article. Hentz's other two publications in 1842–50 and 1875 (the latter posthumously published) somehow became the de facto authority in the United States, earning him the moniker "father of American Arachneology."[42]

However, in 1888 Henry Christopher McCook (1837–1911), an American Presbyterian minister and naturalist who wrote many books on natural history, including spiders, delivered a paper at the Academy of Natural Sciences in Philadelphia, arguing that with the exception of the 125 species Hentz had published in 1832, Walckenaer's names were the correct ones to use by right of priority, since the baron had published these names first, four years before Hentz. McCook acknowledged that to change from Hentz's to Walckenaer's names would be an inconvenience, creating "a peculiar hardship to American araneologists," but he deemed it was the proper thing to do, since Walckenaer had published these names first.[43] Some arachnologists in the 1880s were convinced by this argument that Walckenaer's descriptions and names published before Hentz's should supplant Hentz's, while others argued that with the forty intervening years of studying spiders using Hentz's names, the proposed name changes would throw the field into a state of confusion. If McCook's proposition had succeeded, not only would Walckenaer's reputation have been enhanced and his name forever attached to the hundreds of arachnids he named first, but also Abbot's authority as an expert on the natural history of spiders would have been more visible as the acknowledged source for Walckenaer's descriptions of these new species.

A contemporary critic of McCook's proposition protested that "Mr. McCook is inclined, however, to set too high a value on these drawings."[44] McCook's assessment of Abbot's spider drawings was, indeed, quite positive. He had described them as "tolerably accurate, well finished," and "colored after nature," and "there was no difficulty at first sight in identifying a large number of our well known species, under the names published by Professor Hentz." He concluded that Abbot should have had the "distinction of being the father of American araneology," but his "inability to publish or procure the publication of his faithful labors" robbed him of that distinction. McCook thought adopting Walckenaer's names would be a way to render Abbot "justice and honor."[45] Another way to give Abbot his due was to publish his spider drawings, a suggestion that McCook made and that was seconded by an eminent Swedish arachnologist, Professor Tamerlan Thorell (1830–1901), who contributed these encouraging words to this debate: "Would it not be possible to have Abbot's work published? There are in America so many wealthy citizens who are willing to make sacrifices for scientific purposes; and in this case an appeal to the national feeling of your countrymen, would not, I think be out of place."[46] Unfortunately for Abbot's reputation, McCook's and Thorell's hopes of publication have yet to be realized, and Chamberlin and Ivie's assertion that Abbot's original drawings "surpass in accuracy as well as artistry the drawings of Hentz" has gone unheeded.[47] As a result, Abbot's contribution to the study of North American spiders suffers to this day from a lack of recognition.

McCook's efforts in the late 1880s to bestow "justice and honor" on Abbot did not work out as he had hoped, and by the second decade of the twentieth century, Dow's perception of Abbot as lacking proper scientific qualifications had prevailed. It's time to cast aside Dow's aspersions and to recognize Abbot's extraordinary contribution as an artist and a naturalist to the study of the North American natural world, filled (as it once was) with exquisitely beautiful butterflies and moths, jewel-like beetles, gossamer-winged dragonflies, elegant wasps, busy bumblebees, and nimble spiders. It is my hope that the fascination Abbot felt for these creatures can inspire us today so that we may learn to love insects and spiders just a bit more than most of us do—for their sakes and ours.[48]

Notes

Preface and Acknowledgments

1. See, for example, Fascicle 5.1 on Sesiidae by Thomas D. Eichlin and W. Donald Duckworth published in the series *The Moths of America North of Mexico* (1988).

Chapter 1. "A Peculiar Liking for Insects"

1. Andrey Avinoff, quoted in Wilkinson, "John Abbot's London Years. Part 1" (1984), 117. "A peculiar liking" is Abbot's phrase from his very brief autobiography, which has been published in Gilbert, *John Abbot* (1998).
2. Andrey Avinoff, transcript of January 1946 letter, Andrey Avinoff Papers, Archives of the Carnegie Museum of Natural History, Pittsburgh, Pa. See also Geoffrey T. Hellman's "Black Tie and Cyanide Jar" (August 13, 1948), a *New Yorker* profile of Avinoff, who had recently retired, celebrating Avinoff's career as a lepidopterist, museum curator, and administrator: "With seven thousand items in his library to choose from, he selected an album about insects to carry with him—a late eighteenth-century volume, never published, of a hundred watercolor plates of unusual beauty and accuracy, painted by John Abbot, an American lepidopterist" (39).
3. Abbot is credited with conducting the first regional inventory of North American insect and bird life. See Allen, "History of American Ornithology Before Audubon" (1951). For other eighteenth-century artist naturalists who drew the natural world of the American South, see Magee, *The Art and Science of William Bartram* (2007); and McBurney, *Illuminating Natural History* (2021).
4. Abbot's drawings and observations of birds are not covered in this book. Alexander Wilson was quite impressed with Abbot's knowledge of bird life and sought him out, going birding with him in Georgia. For more information about Abbot's knowledge of birds, see Simpson, "The Artist-Naturalist John Abbot (1751–ca. 1840): Contributions to the Ornithology of the Southeastern United States" (1984); Simpson, "Artistic Sources for John Abbot's Watercolour Drawings of American Birds" (1993); Gilbert, *John Abbot*; Rogers-Price, *John Abbot in Georgia* (1983).
5. James Edward Smith to Richard Pulteney, 1 May 1797, GB-110/JES/MS238/17, Linnean Society, London.
6. The quotation is from Swainson, *Taxidermy* (1840), 100. The most comprehensive treatment of Abbot to date is Gilbert's *John Abbot*. See also Wilkinson, "Smith and Abbot, *The Natural History of the Rarer Lepidopterous Insects of Georgia* (1797), and Its Authorship and Later History" (1981); and Wilkinson, "John Abbot's London Years. Parts 1–4." For careful scrutiny of the publication process of Abbot and Smith's volume, see Calhoun, "A Glimpse into a 'Flora et Entomologica'" (2006). For Abbot's involvement with other naturalists, see Calhoun, "*Histoire générale et iconographie des lépidoptères et des chenilles de l'Amérique septentrionale*" (2004); Calhoun, "John Abbot's 'Lost' Drawings for John E. Le Conte in the American Philosophical Society Library, Philadelphia" (2006); and Calhoun, "John Abbot's Butterfly Drawings for William Swainson" (2007). Art historical treatment of Abbot is negligible, with the important exception of the book by Neri, Nummedal, and Calhoun, *John Abbot & William Swainson* (2019); and Rogers-Price's groundbreaking exhibition catalog, *John Abbot in Georgia*. For the history of entomological art in early modern Europe, see Neri, *The Insect and the Image* (2011).
7. Abbot, "Notes on My Life," in Gilbert, *John Abbot*, 118.
8. Georgia was the last and least settled of the thirteen colonies, with European settlement confined mostly to the seacoast and the Savannah River basin. The territory between the coast and the Creek and Cherokee nations was still very much of a frontier when Abbot arrived in 1776. For a history of the Creek Wars and the removal of Muscogee and Cherokee from Georgia in the 1830s, see, respectively, Saunt, *A New Order of Things* (1999); and Ehle, *Trail of Tears* (1989).
9. Maria Sybilla Merian went to Surinam, Henry Smeathman to Sierra Leone, and John William Lewin to Australia.
10. For Philadelphia as the center of American natural history inquiry, see Meyers, *Knowing Nature* (2011).
11. What is known of Abbot's private life is derived mostly from legal documents concerning the transfer of mobile property, which included human beings. In his will he

left all his belongings, including his artwork, scientific materials, and Betsey, an enslaved woman in her forties, to his friend and neighbor William E. McElveen (Abbot's will, October 24, 1839, Record Book 5 [1829–40], Bulloch County Courthouse, Statesboro, Ga.). Freeing slaves was illegal in Georgia at the time, and such transfers of ownership might have been regarded in that era as a kindness to enslaved people, allowing them to remain in a place they had resided in for some time, an act that might have afforded them some small solace.

12. Jean-Jacques Rousseau, *Reveries of a Solitary Walker* (1782), 77, quoted in Marder and Vieira, "Writing Phytophilia" (2013), 40.
13. Drury, preface to *Illustrations of Natural History* (1770), 1:vii.
14. See "A Catalogue of the Exotic Insects in the Collection of Dru Drury, 1784," Hope Entomological Collections, Library and Archives, Oxford University Museum of Natural History, Oxford, U.K. For a theory of wonder and a refusal to equate science with disenchantment, see Kareem, *Eighteenth-Century Fiction and the Reinvention of Wonder* (2014); and Chico, "From Fleece to Fleets" (2019).
15. Drury, *Illustrations of Natural History*, 1:vi.
16. Drury, 1:vi, vii. This excitement over the intricate workings of insect bodies did not originate in the late eighteenth century. Seventeenth-century natural philosophers such as Jan Swammerdam delighted in detailing insect bodies as elegant machines that reflect the divine creation of the natural world. For Swammerdam's enthusiasm for "the humblest of creatures" by which "divine wisdom and supreme skill are made known," see Jorink, "Between Emblematics and the 'Argument from Design'" (2007), 147. For Robert Hooke's enthusiastic appraisal of "the intricate assemblage of interlocking parts that make up the insect's mechanism" as "a miracle of design," see Festa, *Fiction Without Humanity* (2019), 126. For early modern discourses that figure insects as pests, parasites, and vectors of disease, see Cole, *Imperfect Creatures* (2016).
17. Bullock, *Companion to the Liverpool Museum* (1805), 58–59.
18. Kirby and Spence, *Introduction to Entomology* (1816–26), 1:16–17.
19. This argument from divine design was pervasive in Enlightenment Europe, with Jan Swammerdam as an early proponent. See Ogilvie, "Nature's Bible" (2008). For the culture of early eighteenth-century British science, see Festa, *Fiction Without Humanity*; Chico, *The Experimental Imagination* (2018); and Chico, "Gimcrack's Legacy" (2008).
20. Abbot, "Notes and Observations on the Drawings of the Spiders of Georgia," vol. 14, [fol. 2], Francillon Collection, Entomology Library, Natural History Museum, London (NHM). Versions of this passage reappear in the notes that accompanied his drawings of the dirt dauber. Drury's transcription of this note is written on the back on Abbot's dirt dauber drawing, now in Harvard's Houghton Library.
21. Abbot, "Drawings of the Insects of Georgia," vol. 1, fols. 8, 38, vol. 4, fol. 40, and vol. 5, fol. 38, Francillon Collection.
22. Bergson, *Creative Evolution* (1998), 140, quoted in Parikka, *Insect Media* (2010), 20. For fitness, see Daston, "Attention and the Value of Nature in the Enlightenment" (2003), 123.
23. For poetry about insects, see, for instance, Anna Letitia Barbauld, "To Mrs. P.," and Charlotte Smith, "To the Insect of the Gossamer," both discussed in my introduction to "Critical Insect Studies," a special issue of *Eighteenth Century: Theory and Interpretation* (2024).
24. For natural history's entanglement with empire and slavery, see, for instance, Murphy, "James Petiver's 'Kind Friends' and 'Curious Persons' in the Atlantic World" (2020); Murphy, *Captivity's Collections* (2023); Coleman, *Henry Smeathman, Flycatcher* (2018); and Delbourgo, *Collecting the World* (2017). For natural history and the Spanish Empire, see Bleichmar, "Painting as Exploration" (2006); and Bleichmar, *Visual Voyages* (2017).
25. Abbot had a financial relationship with a Savannah cotton factor, William Gaston, who acted as a kind of banker for Abbot, receiving payments from English collectors, which he would pass on to Abbot. Swainson's correspondence at the Linnean Society contains directions from Abbot about how Swainson was to remunerate him:

> Savannah 1st. May 1818
> I have taken the Liberty to draw a Bill on You for 103 Dollars ["£23 sterling" added presumably by Swainson] in favor of M^{r} Wm. Gaston Merchant in Savannah, Please to pay the balance of my Account whatever may be to M^{r} Wm. Hopkins Correspondent in Liverpool for me, to receive of Mr Gaston Savannah. . . . Direct for me to the care of Mr. Wm. Gaston, merchant, Savannah:
> To 900 insects at 5^{d} each £18.15^{s}
> To 104 Drawings
> To Large corked box

For the history of cotton as a global commodity, see Beckert, *Empire of Cotton* (2014).

26. Abbot to Swainson, 1 May 1818, Savannah, Swainson Correspondence, Linnean Society, London.
27. The document that contains this information about the breakup of Abbot's household is reproduced in Bassett, "Some Georgia Records of John Abbot, Naturalist" (1938), 248. Abbot may have transferred ownership of enslaved workers to his daughter-in-law, Eliza, who had separated from her husband as a way to protect her (and them) from the consequences of his son's bankruptcy and abusive behavior (248). Abbot's son died in his forties in 1826 from a "liver complaint" (250). However, some

of Bassett's assumptions are questionable, most notably, her assertion that Abbot fought for the American rebels in the Revolutionary War and that he received a large land grant as payment for military service. Wilkinson, in "John Abbot's London Years. Part 4," 284, argues convincingly that the person named John Abbot that Bassett found mentioned in military records could not have been our John Abbot because the person in these records was illiterate, signing his name with an X. (Wilkinson credits Elsa G. Allen for this discovery.)

28. The deed documenting the transfer of property from Abbot to his daughter-in-law describes how enslavement robbed people of the fruit of their labor. Eliza G. Abbot was permitted "to receive and take all Sum or Sums of money earned by them [the enslaved] and all profits of what nature soever accruing or to accrue from their labor from henceforth and for her own use and benefit forever" (Bassett, "Some Georgia Records," 248). Whatever money these enslaved workers might have earned as skilled laborers or talented crafters "belonged" to their enslavers.

29. Neri wonders if Abbot "may have had help in collecting his specimens and preparing his drawings" and if these "assistants" might have been enslaved people (*Insect and the Image*, 15). "Assistants" is Swainson's term in the Abbot entry in *Taxidermy*, in which he suggests that Abbot's drawings of the 1820s and 1830s were inferior to his earlier work and that his use of assistants may have resulted in this lesser work. Swainson may have been referencing the book of drawings Abbot had sent him in the mid-1830s, which made its way back to the United States via Edmund Doubleday. See chapter 4 for details about these drawings. Calhoun disagrees with Neri's supposition that enslaved people may have been responsible for these "inferior" drawings. See Calhoun, "Butterfly Drawings by John Abbot in the Houghton Library, Harvard University, That Are Wrongly Attributed to an 'Inferior Copyist'" (2007).

30. The following passage is from a letter written in April 1834 by Augustus Oemler to Thaddeus Harris, who had inquired about Abbot's current condition: "Mr. A. used glasses ever since I have known him (1805) and, by means of them, paints the smallest Insects with pretty much correctness, but his hearing faculties are much impaired, the last time I saw him (1825) I had to use the Slate to converse with him. But not withstanding this misfortune, he was chearful, and his constitutional smile never left his countenance. He is now, I am told, very corpulent, but still exercises his pursuit of hunting birds and drawing—but engages boys to run after Butterflies." This letter was published by Dow in 1914 in Dow, "John Abbot, of Georgia," 70. A family tradition has it that one of these boys who caught butterflies for the very old Abbot was Aaron Cone, whose grandson Howell C. Cone informed Anna Stowell Bassett of these details. See Bassett, "Some Georgia Records," 253. Titian R. Peale apparently told W. H. Edwards in 1864 that Peale had visited Abbot when the latter was living in Savannah; Peale said that "negro boys brought" Abbot "caterpillars of all sorts." See Dos Passos, "The Entomological Reminiscences of William Henry Edwards" (1951), 149–50.

31. For more on the intertwined histories of museums, natural history collecting, and empire, see Delbourgo, *Collecting the World*; Endersby, *Imperial Nature* (2008); Findlen, *Possessing Nature* (1994); Goodman, *Planting the World* (2021); Impey and MacGregor, *The Origins of Museums* (1985); MacGregor, *Curiosity and Enlightenment* (2007); and Miller and Reill, *Visions of Empire* (1996). For studies of collecting as a social practice, see Elsner and Cardinal, *The Cultures of Collecting* (1994); and Pomian, *Collectors and Curiosities* (1987).

32. For a defense of the scientific importance of natural history museums' collections and their role in preserving biodiversity, see Kemp, *The Lost Species* (2017).

33. For a thorough account of the whereabouts of Abbot's butterfly specimens and his drawings, see Calhoun's excellent article "From Oak Woods and Swamps" (2019). I am grateful to Isabelle Charmantier, head of collections at the Linnean Society, for showing me the Lepidoptera that had labels with "Georgia" or "Georgia Abbot" attached to them on February 20, 2019.

34. Calhoun, "From Oak Woods and Swamps."

35. Swainson became a proponent of quinarianism, a system of biological classification based on the number 5, which made him unpopular with the zoological establishment.

36. Swainson, *Taxidermy*, 99–100.

37. Much to the consternation of entomologists and other like-minded individuals, I will, in general, use the word "insect" rather than "arthropod" because the word "arthropod," a Latin-derived term not in common parlance, can be offputting to the general reader. ("Insect" is derived from the Latin *insectum*, but the word has been thoroughly Anglicized.) My goal in using the word "insect" is to make these tiny animals less alien and more approachable. Technically, spiders are not considered insects; they are arachnids that also belong to the phylum Arthropoda.

Chapter 2. "Going Abroad to Collect Foreign Insects"

1. Abbot, note x235, "Aranea" (ca. 1799), fol. 25, Rare Book Folio QL458.42.A7 A2 1799, Hargrett Rare Book and Manuscript Library, University of Georgia Libraries. This spider volume and its notes were once owned by the Chetham Library, Manchester, Great Britain; perhaps this institution was responsible for the transcription and binding of Abbot's notes. For more information on Abbot's spider drawings and their afterlives, see Calhoun, "A Persistent Case of Mistaken Identity" (2019). Spider +235 is probably a wolf spider; see figure 7.3.

2. Abbot, note ++90, "Aranea" (ca. 1799), fol. 10, Hargrett Library.
3. The notes accompanying Hargrett Library's "Aranea" are not in Abbot's handwriting, but they have been copied from his original notes.
4. Late in life, Abbot wrote a few pages of an autobiography, "Notes on My Life," which was published in Gilbert, *John Abbot*, 117–24. The identity of Van Dest is uncertain. Abbot may have misremembered the name and may have been referring to Georg Dionysius Ehret (1708–70), a botanical artist born in Germany who had settled in London. Ehret, one of the great botanical artists of his day, drew plants for Britain's natural history elites, including Margaret Cavendish Bentinck, Duchess of Portland, Sir Joseph Banks, and Peter Collinson.
5. Abbot, "Notes on My Life," 117. Bonneau, a watercolorist, is described as the son of a Huguenot engraver who worked in the London book trade and as "one of the drawing masters to the Royal Family, who, although his reputation as a painter was small, received some support when a candidate for Associateship of the Royal Academy" in the 1770s. He ultimately failed to acquire this position, although he received letters of support from Joseph Highmore and Joshua Reynolds. See Whitley, *Artists and Their Friends in England* (1928), 2:316–18.
6. Abbot, "Notes on My Life," 117.
7. Jardin, "Memoir of Dru Drury" (1842), 15:17.
8. Abbot, "Notes on My Life," 118.
9. Graves, *The Society of Artists of Britain* (1907), 1. Poland Street's residents included artists, engravers, architects, and draftsmen. Among them, with their dates of residence, were Charles Burney, musicologist (1760–70), and his family (Frances Burney, who eventually became a novelist, was a child at this time); Paul Sandby, watercolor artist (1767–72); Thomas Sandby, architect (1760–65); Gavin Hamilton, painter (1772); Thomas Malton, architectural draftsman (1772–80); and William Blake, artist, poet, and engraver (1785–91). For the history of Poland Street, see *Survey of London*, vols. 31 and 32, *St. James Westminster*, pt. 2, available on British History Online, https://www.british-history.ac.uk/survey-london/vols31-2/pt2/pp243-249.
10. For Cavendish and Fabricius, see Jardine, "Memoir of Dru Drury" (1842), 44–46, 58–59. See also Wilkinson, "John Abbot's London Years. Part 2."
11. For more on Banks's career, see Gascoigne, *Science in the Service of Empire* (1998); David Phillip Miller, "Joseph Banks, Empire, and 'Centers of Calculation' in Late Hanoverian London" (1996); Werrett, "Rethinking Joseph Banks" (2019); and Goodman, *Planting the World*.
12. Abbot, "Notes on My Life," 118. For Smeathman's activities as a naturalist and specimen collector, see Coleman, *Henry Smeathman, Flycatcher*.
13. For the few details that are known about Abbot's ties to Thomas Martyn, see Wilkinson, "John Abbot's London Years. Part 4." In addition to being a natural history dealer of some prominence, Martyn was also an artist and publisher of illustrated natural history books. In 1797 Martyn published illustrations based on specimens Abbot had collected in a slight volume titled *Psyche: Figures of Nondescript Lepidopterous Insects from Different Parts of the World*. Only twenty-eight copies are extant; these volumes contain very little text and have varying numbers of plates, ranging from two to thirty-two. For details, see Griffin and Swann, "A Revised Census of the Known Copies of T. Martyn's Psyche 1797" (1985). For Martyn's Academy for Illustrating and Painting, see Tobin, "Butterflies, Spiders, and Shells" (2012).
14. For an overview of Abbot's time in Virginia, see Calhoun and Johnson, "John Abbot, the English Naturalist-Artist, in Virginia" (2010).
15. For biographies of these men, see Wyatt, "Dr. James Greenway" (1981); and Jarratt, *The Life of the Reverend Devereux Jarratt* (1806).
16. Drury to Devereux Jarratt, London, 30 September 1766, Drury's Letter-book, 1761–83, SB f D.6, Entomology Special Collections, NHM.
17. Drury to Abbot, 10 April 1774, Drury's Letter-book.
18. Drury to Dr. James Greenway, 5 May 1773, Drury's Letter-book.
19. Drury mentioned Abbot's industriousness in Drury to Smeathman, 28 November 1774, Drury's Letter-book. The quote is from Abbot, "Notes on My Life," 122.
20. Drury to Abbot, 10 April 1774, Drury's Letter-book.
21. Abbot, "Notes on My Life," 122.
22. Drury to Abbot, 10 April 1774, Drury's Letter-book.
23. Abbot, "Notes on My Life," 124.
24. Rubens Peale to Charles Peale, 23 September 1802, quoted in Lillian B. Miller, *The Selected Papers of Charles Willson Peale* (1988), 2:458. This is the only mention in the historical record of Abbot being a schoolteacher, which, considering the source, is open to question. See below Abbot's condemnation of war between Britain and the United States, as it threatened his livelihood as an artist and specimen hunter.

    Rubens Peale's account of Francillon's collection is important because it registers how impressed he is with it, but seven thousand insects is a low figure if it is compared to the number of insects that appear in the Francillon auction catalog in 1816; seven thousand may refer to the number of American specimens. I also doubt that Francillon's collection was for sale. The 2,000 pounds sterling may have been what it was worth, which is not the same as it being for sale. Francillon did, however, broker sales of Abbot's specimens and drawings, and this may be what led Rubens to assume the whole collection was for sale.
25. See Gilbert, *John Abbot*, 64.
26. Abbot to Escher, 18 April 1813, John Abbot Letters to Henry Escher, 1813–1836, Archives 4600-0004, Rare and Manuscript Collections, Cornell University Library. According to Calhoun, "In 1858, Escher-Zollikofer's collection of over 60,000 specimens was donated by Alfred Scher (his son) and Kaspar Stockar-Escher (husband of

daughter Clementine) to the Eidgenössische Technische Hochschule (Federal Polytechnic School) in Zürich" ("From Oak Woods and Swamps," 227).

27. See Calhoun for details about numbers of insect specimens Abbot sent him, "at least 6768 insects," of which 2,331 were Lepidoptera ("From Oak Woods and Swamps," 227). Calhoun calculates that thousands of these specimens are in the Eidgenössische Technische Hochschule Zürich (Swiss Federal Institute of Technology), "making it perhaps the largest known repository of his material" (227).
28. Swainson to Abbot, 25 October 1817, Bahia Brasil, Swainson Correspondence.
29. Swainson to Abbot, 26 February 1818, Bahia Brasil, Swainson Correspondence.
30. Swainson, *Taxidermy*, 100.
31. Abbot had sent him in 1818 "104 Q[quarto] Drawings of the changes of the insects of Georgia, making a 2d. Vol. of Smith" (Abbot to Swainson, 1 May 1818, Savannah, Swainson Correspondence).
32. For an excellent overview of Abbot and Swainson's (failed?) collaboration, see Neri, Nummedal, and Calhoun, *John Abbot & William Swainson*.
33. Abbot to Swainson, 1 May 1818, Savannah, Swainson Correspondence.
34. For the Macleay Museum's eighteenth-century insects, see Coleman and Blackburn, "Eighteenth-Century West African Insects in the Macleay Museum, University of Sydney" (2017).
35. For a photograph of this red-winged horse lubber grasshopper (*Taeniopoda eques*) that was collected by Abbot, see Stacey and Hay, *Museum* (2007), 163.
36. *A Catalogue of the Extensive Collection of Insects . . . from the Museum of George Milne, F.L.S.* (1824). In the 1830s Edward Doubleday wrote to his friend Thaddeus W. Harris: "In all old collections are many specimens collected by Abbot; at Francillon's, Donovan's and other sales, some of these have been dispersed, and have crept into collections nominally British only. At Milne's sale, a large number of Abbot's insects were bought by the British Museum. The large Sphinx I mentioned in Abbot's drawing, of which one was sold at his sale, is *Labruscæ*" (Scudder, *Entomological Correspondence of Thaddeus William Harris* [1869], 147).
37. The LeContes, a family with Georgia, New Jersey, and Pennsylvania branches, produced several famous nineteenth-century scientists, including New Yorker John Lawrence LeConte (1825–83), who was considered the premiere American entomologist of his era. His Southern cousins were John LeConte (1818–91), physicist and president of the University of California, Berkeley, and Joseph LeConte (1823–1901), a professor of geology at the University of California, Berkeley, a founding member of the Sierra Club, and the author of white supremacist tracts.
38. Abbot to Swainson, 20 December 1816, Savannah, Swainson Correspondence.
39. Louis LeConte had a botanic garden on his Woodmanston plantation. Spencer Fullerton Baird, the second secretary of the Smithsonian, possessed hundreds of Abbot's bird drawings, and I assume that he acquired them through the LeContes, since Baird was J. E. LeConte's distant cousin, and LeConte's son, John Lawrence LeConte, was good friends with Baird as well as a colleague. Their correspondence is housed in the Archives of the Smithsonian Library.
40. Abbot, plate 14, vol. 12, fol. 69, Francillon Collection. For a drawing of this black wasp with three long streamers, see Gilbert, *John Abbot*, 106.
41. Abbot's description copied by Drury, fol. 41v, MS Typ 426.2, Houghton Library, Harvard University.
42. Abbot, "Pine Cerambyx," "Drawings of Insects and Plants," sequence 59–60, Houghton Library, Harvard University. This beetle is also figured in "Moths of Georgia," vol. 1, plate 2, as well as in the Francillon Collection's vol. 2 (1792), plate 249, and vol. 16 (1804), plate 75, with similar notes.
43. Abbot, "Pine Cerambyx," TAB. LXXV, vol. 16, Francillon Collection. J. K. Wadleigh explained how she went about identifying this species in Abbot's drawing:

> Definitely one of the pine sawyer species; there's 3 long-horned beetles that are likely for it to be in Georgia. *M. notatus* and *M. carolinensis* are the other possible identifications. For *notatus*, it tends to be darker overall and less contrasted in its pattern—giving it a "washed-out" gray look. When it does have more of a pattern it's divided up with distinct vertical stripes down the elytra that are darker grey and patternless. The spotting is also smaller. *M. carolinensis* is the closest contender, but still has less of a prominent pattern than *titillator*. *Titillator* is very contrasted and the pattern matches well. It also looks like he drew the detailed curved and pointed tips to the elytra (called elytral spines) that is a notable feature in this species. (personal communication, February 2025)

44. Jardine, "Memoir of Dru Drury," 49–50.
45. On the swallowtail caterpillar, see Abbot, "Papilio Ajax," TAB. LXXVIII, vol. 17, drawing 20, Francillon Collection; on the red wasp, see Abbot, "Papilio Antiopa," TAB. V, vol. 17, drawing 26, Francillon Collection.
46. Abbot, note +229, "Aranea," sec. 2, fol. 25, Hargrett Library.
47. Abbot, note +233, "Aranea," sec. 2, fol. 25, Hargrett Library.
48. Abbot, note +210, "Aranea," sec. 2, fol. 22, Hargrett Library.
49. Abbot, note 33, "Aranea," sec. 1, fol. 4, Hargrett Library.
50. Abbot, "Sphex," vol. 13, drawing 191, Francillon Collection.
51. Abbot, *Sphex sabulosa*, "Drawings of Insects and Plants" (ca. 1786), MS Typ 426.2. This wasp is also pictured in vol. 13, drawing 195, Francillon Collection. Abbot gave the number 56 to the wasp in both of these drawings, a prac-

tice that enabled him to keep track of what arthropod he had drawn and distributed to his clients. The name *Sphex sabulosa* refers to the red-banded sand wasp (*Ammophila sabulosa*), but this wasp's distribution is Europe and northern Asia (excluding China).

52. Kirby, "Ammophila, a new Genus of Insects in the Class Hymenoptera" (1798).
53. The spider drawings held in the Hargrett were drawn on paper with a watermark "J. Whatman. 1794," and the bound manuscript notebook is on laid paper with the watermark date of 1799. The notebook contains three sections: Section 1 has numbers running 1 to 80; section 2 has numbers running +111 to +318; and section 3 runs ++1 to ++250. These notes match the drawings and their numbers, totaling 457 entries and 457 drawings. There is a gap in numbered notes, with section 2 beginning with page 111, while the other two sections begin with page 1. By my count, there are three sets of spider drawings plus their notes: Hargrett's; Francillon's volume 14 in NHM's Entomology Library; and Walckenaer's in the library of the Muséum national d'Histoire naturelle, Paris.
54. Abbot, note ++68, "Aranea," sec. 3, fol. 8, Hargrett Library. J. K. Wadleigh described the characteristics of this spider: "This species is very tiny and also extremely variable. The amount of greenish-spotting and red with the white dorsal band varies between individuals, but with the pointed triangular-shaped abdomen and the white spotting pattern it's pretty unique" (personal communication, February 2025).
55. Abbot, note ++137, "Aranea," sec. 3, fol. 17, Hargrett Library.
56. Abbot, note ++14, "Aranea," sec. 3, fol. 2, Hargrett Library. This passage is quoted by Walckenaer, *Histoire naturelle des insectes* (1837–47), 1:248; this is one of eighty-nine citations in the first of four volumes. See chapter 7 for a discussion of Walckenaer's acquisition of Abbot's spider drawings and notes.
57. Abbot, note 3, "Aranea," sec. 1, fol. 1, Hargrett Library.
58. Abbot, note ++25, "Aranea," sec. 3, fol. 3, Hargrett Library.
59. Abbot, note 23, "Aranea," sec. 1, fols. 3–4, Hargrett Library.
60. Abbot, no. 16, "The Imperial, or Great Persimmon Moth," in "A Natural History of North American Insects," Smith MSS vol. 39, "Varia Scientific," Linnean Society, London. (This is the manuscript that became Smith and Abbot's book *Insects of Georgia.*) For Smith's edited version of this description, see Smith and Abbot, *Insects of Georgia*, 2:121, TAB. LXI. For the proper name of this bombyx and a similar description and drawing found in Francillon's collection (vol. 16, fol. 86), see Gilbert, *John Abbot*, 84.
61. Abbot, "A Natural History of North American Insects," 8, Smith MSS vol. 39, "Varia Scientific."
62. Abbot, "Acarus," vol. 15, fol. 10, Francillon Collection.
63. Abbot, "Musketoes," vol. 13, drawings 17, 18, 19, Francillon Collection.
64. For the dispossession of Native Americans in the South, see Saunt, *Unworthy Republic* (2020).
65. For the early history of the lumber industry in Georgia, see Pressly, *On the Caribbean Rim* (2013).
66. Abbot to Escher, 18 April 1813, Savannah, Abbot Letters to Escher.
67. Abbot to Swainson, 10 November 1818, Bulloch County, Swainson Correspondence. The War of 1812 ended in December 1814.
68. Abbot, note 64, "Notes to the Drawings of Insects," fol. 7, "Moths of Georgia," Hargrett Library.
69. Abbot, note 73, fol. 8, "Moths of Georgia," Hargrett Library.
70. Abbot, "Tobacco Bud Worm Moth," no. 96, "A Natural History of North American Insects," Smith mss vols.39, "Varia Scientific."
71. Abbot, note 74, fol. 8, "Moths of Georgia," Hargrett Library.
72. For mixed crops as a way to deter insect infestation, see Sadia Afrin et al., "Intercropping Empower Reduces Insect Pests and Increases Biodiversity in Agro-Ecosystem" (2017). See also https://learn.uvm.edu/foodsystemsblog/2014/01/09/controlling-pests-with-plants-the-power-of-intercropping/. William Leach argues in *Butterfly People* (2013) that family farms of the nineteenth century, "while they sacrificed virgin forests and ecosystems in the short term, contributed over the long term to nature's vitality. Their distinguishing features were not just plowed fields or barns or silos but also ponds, woodlots, hedgerows, stone walls, open fields along roadsides, and meadows by streams or riverbeds for grazing cattle, all created for human purposes but also serving as likely habitats and hideouts for animals" (xvii). Leach argues that his model of the family farm is not to be confused with "extensive monocultural forms" of farming, which "led to the eradication of diverse habitats, not to their creation, and, thereby, to the extinction or near extinction of many butterflies" (292).
73. Evidence that he raised these moths is in his notes: The ground worm moth "went into the ground to change 20th May. bred June 2^{d}," and the "Cotton Caterpillar Moth" "spun up in a leaf 24th August. bred 3^{d} Sepr." (note 74, fol. 8, "Moths of Georgia," Hargrett Library). This album of drawings is very similar to Swainson's album in New Zealand.
74. Abbot's manuscript for *Insects of Georgia*, 13. See also plate 45 in *Insects of Georgia*, 1:89, and plate 24 in the Francillon Collection, vol. 6, fol. 88 (in Gilbert, *John Abbot*, 94).
75. For the failed silk industry in the American South, see Stephens, "The Silk Industry in Georgia" (1953); Marsh, "The Republic's New Clothes" (2012); Marsh, "Silk Hopes in Colonial South Carolina" (2012); and Bigelow, "Gendered Language and the Science of Colonial Silk" (2014).
76. For details concerning the publication of Drury's book,

see Noblett, "Dru Drury's Letters (1770–1775) to the Cambridge Bookseller, John Woodyer" (1994).

77. Jardine, "Memoir of Dru Drury," 32.
78. Naturalist artist Edward Donovan gives a detailed description of the two kinds of nets Drury sent to his collectors. First, the batfowling net (Drury calls it a "batfolding net") was four or five feet in length, and the frame was made out of cane covered with "Scotch gause." It was used "to catch moths on the wing" and could be used like a beating sheet to catch "not only a number of Caterpillars . . . but many of the minuter kinds of coleopterous, and other Insects also." Second, the forceps net was "about ten or twelve inches in length," and its frame was made from steel covered with fine gauze. Its handles "are held and moved as a pair of scissars, and where they can be conveniently used are to be preferred" (Donovan, *Instructions for Collecting and Preserving Various Subjects of Natural History* [1794, 1805], 22–23).
79. For the intersection of natural history collecting with the British slave trade, see Douglas, "The Making of Scientific Knowledge in an Age of Slavery" (2008); Coleman, *Henry Smeathman, Flycatcher*; and Murphy, *Captivity's Collections*, especially chapter 5, "Searching for Goliath," 125–51.
80. Drury, *Directions for Collecting Insects in Foreign Countries*, in Jardine, "Memoir of Dru Drury," 33–37.
81. Drury, *Illustrations of Natural History* (1782), 3:iv.
82. Drury to Captain Mayle, reproduced in Jardine, "Memoir of Dru Drury," 29–30.
83. Jardine, 49.
84. For Drury and Smeathman's relationship, see Coleman, *Henry Smeathman, Flycatcher*; and Murphy, *Captivity's Collections*.
85. Smeathman's observations about African insects were incorporated into Drury's third volume.
86. Woodward and Petiver, quoted in Delbourgo, *Collecting the World*, 97–98. For more on Petiver, see Murphy, "James Petiver's 'Kind Friends' and 'Curious Persons' in the Atlantic World" (2020).
87. James Petiver to Madam Carter, 6 January 1696, Sloane MS 3332, fol. 173v, British Library, quoted in Murphy, "Petiver's 'Kind Friends,'" 273.
88. John William Lewin to Drury, 7 March 1803, Sydney, Separate MSS of Dru Drury, Entomology Special Collections, NHM.
89. For European naturalists' disregard for their colonial counterparts, see Parrish, *American Curiosity* (2006).
90. For Latour's centers of calculation, see *Science in Action* (1987). For examples of how scientific knowledge is produced through abstraction and decontextualization, see Latour, *Pandora's Hope* (1999).
91. Samouelle, *General Directions for Collecting and Preserving Exotic Insects and Crustacea* (1826), 11–12. Coleman, *Henry Smeathman, Flycatcher*, makes the astute observation that the British representation of insects as a "decontextualized and isolated artefact" may have sufficed for entomological books on British insects, since their life histories, habitats, and ecological interactions with their environments would have been known to British readers, but this format was insufficient for non-European insects, about which little was known in British natural history circles (216).
92. Bleichmar, "The Geography of Observation" (2011), 392. See also Tobin, "Imperial Designs" (1996).
93. For colonists as specimen collectors and the hierarchy of knowledge practices, see Parrish, *American Curiosity*.
94. Samouelle, *General Directions for Collecting*, 11.
95. This process has been described by historians of science as "collaborative empiricism" and "distributed or extended cognition." These terms, in stressing the spatial dynamics of the movement of natural knowledge across the globe, are meant to complicate our understanding of empiricism as an exclusively European practice, but they fail, I believe, to register the power differentials at work within such systems. For collective empiricism, see Daston and Galison, *Objectivity* (2007), 19–23; and Schaffer et al., *The Brokered World* (2009). For power differentials within these collective systems, see Bleichmar, "The Geography of Observation"; and Bleichmar, "Paintings as Exploration."

Chapter 3. "The Harvest of Many Years of Industry"

1. The title quotation is from a letter Abbot wrote to Escher, 18 April 1813, Savannah, Abbot Letters to Escher. Calhoun suggests in "From Oak Woods and Swamps" that there are more Abbot-collected specimens than just these eleven in the Linnean insect collections—twenty in all. I concur with his count and took photographs of twenty specimens that have "Georgia" or "Abbot and Georgia" on their tags.
2. Smith and Abbot, *Insects of Georgia*, 1:85.
3. Smith identified the plant pictured in *Insects of Georgia*, TAB. XLIII, as a "*Tabernaemontana Amsonia*," but I think, based on the narrow leaves, that it is probably a ciliate bluestar (*Amsonia ciliata*), neither of which, by the way, seems to be a host plant for this particular clearwing moth (*Hemaris diffinis*).
4. Abbot to Escher, 2 November 1821, Bulloch County, Abbot Letters to Escher.
5. Swainson, *Taxidermy*, 99. Writing shortly after Abbot's death, James Duncan, a Scottish entomologist, explained that Abbot had "transmitted many fine specimens of insects to private individuals in Britain, as well as to public museums both here and on the continent" (*Natural History of Exotic Moths* [1841], 69–70).
6. See Rogers-Price's *John Abbot in Georgia* for the discovery of Abbot's bird specimens in Liverpool (44, 96). See Calhoun, "From Oak Woods and Swamps," for the possibility of hundreds of Abbot specimens extant in Germany and even more in Switzerland.

7. Abbot to Swainson, 20 December 1816, Savannah, Swainson Correspondence. For preserving spiders, see Hancock and Brown, "An Unusual Preparation of an Eighteenth-Century Spider and Its Consequences" (2014).
8. Kirby and Spence, *Introduction to Entomology*, 3:148–49, 4:537–38.
9. He also bred insects as a child but did not know a "method of keeping them after I had done it" (Abbot, "Notes on My Life," 117).
10. Abbot was also a very skillful bird taxidermist and very knowledgeable about birds and their habits, locales, and life histories, so knowledgeable that Alexander Wilson sought him out to go birding together and relied on Abbot to supply him with bird specimens he had not been able to acquire on his own. Abbot was referred to as "a distinguished naturalist" by George Ord, editor of the 1828–29 edition of Wilson's *American Ornithology: Or, The Natural History of the Birds of the United States*, 3:376. Ord asked Abbot for help describing a darter or snake bird (*Plotus anhinga*).
11. For drawings and descriptions of caterpillars he had raised in England before leaving for the American South, see Abbot's manuscript volume titled "A Natural History of Insects. . . . Drawn and coloured from Nature. Together with a concise and accurate Description" (1772), MS Typ 426.1, Houghton Library, Harvard University.
12. Swainson, *Taxidermy*, 14–15.
13. See chapter 4 of this book for a discussion of this genre depicting insects and their host plants.
14. Abbot's manuscript "A Natural History of North American Insects," Smith MSS vol. 39, "Varia Scientific."
15. Donovan, *Instructions for Collecting*, 21.
16. For more on insect-collecting manuals, see Hünniger, "Nets, Labels, and Boards" (2018).
17. Harris, preface to *The Aurelian* (1766). Harris refers to this net as the "bat-folding net," while other writers refer to this kind of net as a "bat-fowling net." This net does fold up, but it could perhaps be used to catch small birds. For more on nets, see Wilkinson, "English Entomological Methods" (1968); and Wilkinson, "The History of the Entomological Clap-Net in Great Britain" (1978).
18. Donovan, *Instructions for Collecting*, 25.
19. Donovan, 22. In eighteenth-century Britain, butterfly enthusiasts adopted the name "Aurelian" to describe themselves and their activities as collectors. The term is derived from *aurelia*, the chrysalis and the supposed golden color it emits just before the butterfly emerges from the chrysalid state. For more information about Aurelians, see Salmon, *The Aurelian Legacy*.
20. Donovan, *Instructions for Collecting*, 23.
21. Samouelle, *General Directions for Collecting*, 66–67.
22. Kirby and Spence, *Introduction to Entomology*, 4:540.
23. Lettsom, *Naturalist's and Traveller's Companion* (1774), 8–9. For more details on building a breeding cage, see Kirby and Spence, *Introduction to Entomology*, 4:515–46; and Swainson, *Taxidermy*, 13–15.
24. Harris, preface to *The English Lepidoptera, or, the Aurelian's Pocket Companion* (1775), xiii.
25. Donovan, *Instructions for Collecting*, 25.
26. Abbot, "Great Brown Emperor Moth," in "A Natural History of North American Insects," no. 10, fol. 13, Smith MSS vol. 39, "Varia Scientific." See Smith and Abbot, "*Phalaena Cecropia*," in Smith and Abbot, *Insects of Georgia*, 1:89; Abbot, "Great Sycamore Moth," in "A Natural History of North American Insects," no. 17, fol. 15, Smith MSS vol. 39, "Varia Scientific." See Smith and Abbot, "*Phalaena Imperatoria* or Great-Plane Tree Moth," in Smith and Abbot, *Insects of Georgia*, 2:109.
27. Abbot, "A Natural History of North American Insects," no. 92, n.p., Smith MSS vol. 39, "Varia Scientific." See Smith and Abbot, "*Phalaena Torrefacta*" (*Apatelodes torrefacta*), in Smith and Abbot, *Insects of Georgia*, 2:151. Now called spotted apatelodes.
28. Abbot, "A Natural History of North American Insects," no. 18, fol. 16, Smith MSS vol. 39, "Varia Scientific." See Smith and Abbot, "Fringe-Tree Hawk-Moth *Sphinx Chionanthi*," in Smith and Abbot, *Insects of Georgia*, 1:67.
29. Abbot to Escher, 2 November 1821, Bulloch County, Abbot Letters to Escher.
30. Abbot to T. W. Harris, 30 August 1835, Bulloch County, quoted in Dow, "John Abbot, of Georgia," 71. If Moses Harris's caution about keeping the dirt moist and cool is right, then maybe Abbot's moth chrysalides died not from cold but from a house that was overheated due to the extremely cold weather.
31. Historians of science have lately argued that scientific activity in this era took place within domestic spaces. See Terrall, "Frogs on the Mantlepiece" (2011); and Terrall, *Catching Nature in the Act* (2014).
32. Abbot, "Cream-barred or Beach Tussock," in "A Natural History of North American Insects," no. 50, fol. 22, Smith MSS vol. 39, "Varia Scientific." Smith's revision reads: "under the loose bark of large Plane trees in swampy places" (Smith and Abbot, *Insects of Georgia*, 2:149).
33. Abbot, "Snail Tussock," plate 96 of "Moths of Georgia" and "Notes to the Drawings of Insects," fol. 12.
34. Abbot, "Description of the Tenth Drawing," figs. 86–87, in "A Natural History of Insects," MS Typ 426.1.
35. Abbot, "Chestnut-bordered Underwing," in vol. 1, TAB. XXXVIII (1792) of two volumes of 166 watercolor drawings, "Original drawings and ms. notes of the insects of Georgia, America." These two volumes, complete with engraved title pages, "Drawings of the Insects of Georgia, in America, by John Abbot of Savannah," were transferred from the Natural History Museum at Tring to the Natural History Museum, London. Volume 1 carries a bookplate of Sir Robert Johnson Eden (1774–1844).
36. Abbot, "A Natural History of North American Insects," no. 82, n.p., Smith MSS vol. 39, "Varia Scientific." See Smith and Abbot, "*Phalaena Pithecium*" (*Phobetron pithecium*), *Insects of Georgia*, 2:147.
37. Abbot, "A Natural History of North American In-

sects," no. 22, fol. 17, Smith MSS vol. 39, "Varia Scientific." Known today as the achemon sphinx. See Smith and Abbot, "*Sphinx Crantor*," *Eumorpha achemon* (Drury 1773), *Insects of Georgia*, 1:81.
38. Abbot, "A Natural History of North American Insects," no. 40, fol. 20, Smith MSS vol. 39, "Varia Scientific." See Smith and Abbot, "*Phalaena Ministra*," *Insects of Georgia*, 2:161.
39. "Winged state" is Smith's term; Abbot often refers to adults as a "Fly." See Smith and Abbot, *Insects of Georgia*, 1:41.
40. Abbot, "A Natural History of North American Insects," no. 2, fol. 11, Smith MSS vol. 39, "Varia Scientific." See Smith and Abbot, "*Papilio Troilus*," *Insects of Georgia*, 1:1.
41. Abbot, "A Natural History of North American Insects," no. 79, n.p., Smith MSS vol. 39, "Varia Scientific." See Smith and Abbot, "*Phalaena Quernaria*," *Insects of Georgia*, 2:205.
42. Abbot, "A Natural History of North American Insects," no. 7, fol. 12, Smith MSS vol. 39, "Varia Scientific." See Smith and Abbot, "*Papilio Proteus*," *Insects of Georgia*, 1:35.
43. Abbot, "A Natural History of North American Insects," no. 47, fol. 22, Smith MSS vol. 39, "Varia Scientific." See Smith and Abbot, "*Phalaena Velleda*" (*Tolype velleda*), *Insects of Georgia*, 2:103. Known as the large tolype today.
44. Harris, preface to *The Aurelian*.
45. For paper technologies, see Müller-Wille and Charmantier, "Lists as Research Technologies" (2012); and Pugliano, "Specimen Lists" (2012). See also Tobin, "A Naturalist's Notebooks" (2016).
46. Harris, *English Lepidoptera*, vii–ix.
47. Harris, vii.
48. Moses Harris, "Hawk Tyger bee," in "Notes and Original Drawings for The Aurelian," 152, Entomology Special Collections, NHM. This breeding journal also contains the work of Moses Harris's son John, who seems to have added a few images and notes of his own. John Harris engraved Abbot's drawings and colored the copy printed on vellum, which is in the University of Georgia's Hargrett Library.
49. Vol. 17, Francillon Collection. Abbot's use of the word "bred" refers to when the butterfly or moth emerged from the chrysalis, not when the butterfly or moth copulated or laid its eggs.
50. Smith and Abbot, *Insects of Georgia*, 1:89.
51. There is significant overlap in drawings and notes between those in Swainson's possession and these two volumes of Francillon's collection. This is also true of the notes that accompanied the drawings in the volume owned by Mr. McDougal, now in the Hargrett Library, University of Georgia.
52. Donovan, *Instructions for Collecting*, 28.
53. How much milkweed could one hundred monarch caterpillars eat? A lot, as a modern-day breeder on YouTube attests. See MrLundScience, "Raising Monarchs—Things to Know First (Help the Monarch Butterfly)," https://www.youtube.com/watch?v=hbK-dEaayrA.
54. Harris and Abbot, I believe, knew each other, or at least knew of each other, as they were members of Drury's circle in the late 1760s and early 1770s.
55. Drury to Captain Mayle of the *Hound* sloop of war, quoted by Jardine, "Memoir of Dru Drury," 28–39. "I apprehend you may for a very trifle hire a boy or a man to collect them [insects] in the fields and woods. . . . If he catches a butterfly or moth, let him give the body a good squeeze, which will kill it, and then stick it in one of the boxes" (29).
56. Donovan, *Instructions for Collecting*, 43.
57. Donovan, 49.
58. Harris, preface to *The Aurelian*, n.p.
59. Donovan, *Instructions for Collecting*, 43–49.
60. For the emergence of anticruelty campaigns in the nineteenth century and their connection with Evangelical Anglicans, see Ritvo, *The Animal Estate* (1989).
61. Donovan, *Instructions for Collecting*, 38.
62. Kirby and Spence, *Introduction to Entomology*, 4:537–38.
63. Kirby and Spence, 4:538.
64. Donovan, *Instructions for Collecting*, 38.
65. Haworth, *Lepidoptera Britannica* (1803), 90–91.
66. Haworth, 91.
67. For a rhetorical analysis of the defense of killing insects for scientific purposes in the early nineteenth century, see Hollerbach, "Of Sangfroid and Sphinx Moths" (1996). For research on insect pain and insect suffering, see Carere and Mather, *The Welfare of Invertebrate Animals* (2019), in particular, Elwood, "Assessing the Potential for Pain in Crustaceans and Other Invertebrates," 147–77. See the Lepidopterists' Society's 2019 "Statement on Collecting," which addresses concerns about overcollecting for commercial purposes (https://www.lepsoc.org/content/statement-collecting).
68. False bottoms to boxes: "I have likewise sent you under the Cork at the bottom of the box, (being a false bottom) 104 Q Drawings of the changes of the Insects of Georgia, making a 2[d]. Vol. to Smith" (Abbot to Swainson, 1 May 1818, Swainson Correspondence).
69. Abbot, notes transcribed by Francillon, vol. 5, fol. 10, Francillon Collection.
70. Darwin to Asa Gray, 22 May 1860, letter 2814, Darwin Correspondence Project, University of Cambridge, http://www.darwinproject.ac.uk/entry-2814.
71. Harris, *The Aurelian*. The letters refer to images on the plate.
72. Abbot, notes on *Dermestes* transcribed by Francillon, vol. 3, plate 103, Francillon Collection.
73. Swainson, *Taxidermy*, 1.
74. Abbot to Swainson, 10 November 1818, Bulloch County, Swainson Correspondence.
75. Abbot to Oemler, 26 September 1833, Bulloch County, Harris Correspondence, Ernst Mayr Library, Museum of Comparative Zoology, Harvard University.
76. Drury to the Honorable Lady Anne Monson, London, 16 March 1774, Drury's Letter-book.
77. Donovan, *Instructions for Collecting*, 53.

78. Harris, *English Lepidoptera*, xiv–xv.
79. See Swainson on pins in *Taxidermy*, 16–17. He warns that a naturalist traveling abroad in tropical climates will need approximately one hundred pins a day for "entomological pursuits," and a variety of types are required.
80. Drury, notebook entry dated December 1801: "Got some long pins for Lewin" (Entomology Special Collections, NHM).
81. Drury to Smeathman, 7 June 1773, London, Drury's Letter-book, fol. 273. For pins, see Hancock, Brown, and Jowett, "Pinned Down" (2011).
82. Drury to Henry Seymer, 20 January 1774, Letter-book, fol. 304.
83. Drury to Henry Seymer, fol. 305.
84. Abbot to Swainson, 16 June 1819, Savannah, Swainson Correspondence. Years later, Abbot complained to Escher about the low quality of Swainson's Brazilian specimens: "They were only threads and patches, I have had no dealings with him since." See 12 November 1836, Bulloch County, Abbot Letters to Escher.
85. For the elder Swainson's activities as a naturalist, see Tobin, *The Duchess's Shells* (2014).
86. "John Ellis, Esq. a gentleman of large property in Jamaica and a particular friend of Mr. Seymer, at his recommendation under took to make as complete a collection of drawings of the natural history of that country as could be procured, and took out with him Mr. Robins of Bath, the first natural history draughtsman of that day at Mr. Seymer's recommendation." The rest of the story is in Nichols, *Literary Anecdotes of the Eighteenth Century* (1814), 8:202.
87. Wilson to Bartram, March 5, 1809, Savannah, quoted in the introduction to Hunter, *Life and Letters of Alexander Wilson* (1983), xcvi. "I will not detain you by a recital of my journey from Charleston to Savannah. In crossing the Savannah river, at the place called the Two Sisters' Ferry, my horse threw himself into the torrent, and had I not, at the risk of my own life, rescued him, would have been drown" (xciii).
88. Abbot, "Notes on My Life," 120; and Abbot to Escher, 12 April 1816, Abbot Letters to Escher.
89. For Audubon and alligators, see Herrick, *Audubon the Naturalist* (1917), 2:22. For Lewin and dysentery, see Neville, *Mr. J. W. Lewin, Painter & Naturalist* (2012), 61.
90. Collins, *An Account of the English Colony in New South Wales* (1798), vol. 1, chap. 16, 191.
91. Collins, 191.
92. John Francillon to John Leigh Philips, 10 August 1793, Add. MS 29533, fols. 80–81, British Library.
93. Francillon to Philips, fols. 81–81v, Francillon's emphasis.
94. Francillon to Philips.
95. For natural history's entanglement with colonialism and slavery, see Murphy, "Petiver's 'Kind Friends.'"
96. Abbot to Ord, quoted in Stone, "Some Unpublished Letters of Alexander Wilson and John Abbot" (1906), 366.
97. Abbot to Escher, 3 September 1813, Savannah, Abbot Letters to Escher.
98. Abbot to Escher.
99. Francillon to Baron Walckenaer, 1802, "Abbot's Georgian Spiders," MS 274, Library of the Muséum national d'Histoire naturelle, Paris.
100. For ships going down with his cargoes of insects, see Abbot, "Notes on My Life," 120–22. See also Parsons and Murphy, "Ecosystems Under Sail" (2012).
101. Abbot directed his clients to write to A. G. Oemler, Savannah apothecary, if they had requests for him. Savannah merchants acted as intermediaries for Abbot, who lived a day's ride from Savannah: "If you would send the Birds to me please enclose my letter in one to my friends (Mr Oemler, Mr Williams, Savannah) informing them by what Ship, you have sent them by, that he may take it for me from on board the Vessel as most probable I shall be in the Country at the Time, and not know of it" (Abbot to Swainson, 15 December 1819, Swainson Correspondence).
102. Abbot to Swainson, 15 January 1820, Swainson Correspondence.
103. Wilson relied on Abbot to provide him with birds from the Southeast, and Catesby enlisted the support of London's natural history collectors, who loaned him specimens to examine for his drawings.
104. Wilson, *American Ornithology* (1808–14), 3:viii.
105. Abbot uses the phrase "not settled yet" to refer to white settlement, ignoring this land's previous tenants, the Creeks, who were violently dispossessed of their lands in southern Georgia and Alabama through a series of wars. See Saunt, *A New Order of Things*.
106. Abbot to Escher, 3 September 1813, Savannah, Abbot Letters to Escher.
107. Abbot to Escher, 31 December 1813, Scriven County, Abbot Letters to Escher. The river Abbot refers to could be the Savannah River, which makes up the eastern border of the county, or the Ogeechee River, which borders the southwestern side of the county. My guess is that Abbot is referring to the Ogeechee River, since this would be new territory for him. He had spent lots of time in Burke County along Brier Creek in the 1790s.
108. Abbot to Escher, 18 April 1813, Savannah, Abbot Letters to Escher. For the number of insect specimens Abbot sent to Escher, see Calhoun, "From Oak Woods and Swamps," 227.

Chapter 4. "An Admirer of Nature's Beauties"

1. The chapter title quote is from Abbot to Swainson, 7 June 1819, Swainson Correspondence.
2. I went to the Library of Congress to examine James Edward Smith's *Exotic Botany* (1806), but after I spent a few hours waiting for its arrival in the reading room, a librarian told me they could not find it. She feared it had been stolen, possibly because it had been owned by Thomas Jefferson and bore his signature. She told me that they had made an appointment for me with the rare book library at Dumbarton Oaks to see its copy of this

book. When I arrived the next day, a librarian ushered me into a special room and sat right across the table from me the entire time I was examining this book, not taking a chance that their copy might disappear too.

3. For a description of "Moths of Georgia" and the butterflies depicted in it, see Calhoun, "Butterfly Drawings in the Hargrett" (2007). Calhoun thinks the annotation is by "J. McDougal," while I've gone back and forth on this, settling finally on "T. McDougal."
4. For how artists made their own watercolor paints, see Tobin, "Learning to Craft" (2020).
5. I have taken a cue from how book history scholarship approaches books and manuscripts as material objects, that is, by examining how their owners and readers interacted with these paper-based objects and by tracking how they were annotated, scribbled on, organized, cut up and recycled, given away, sold, or hoarded. See, for instance, Yale, *Sociable Knowledge* (2016); Yeo, *Notebooks, English Virtuosi, and Early Modern Science* (2014); and Charmantier and Muller-Wille, "Worlds of Paper" (2014).
6. Janice Neri, "Introduction: Science and Sentiment," in Neri, Nummedal, and Calhoun, *John Abbot & William Swainson*, 2. Neri's description of Abbot's drawings for Swainson applies equally to the Hargrett Library's copy of "Moths of Georgia," since they are near duplicates of each other. However, Swainson's drawings were drawn between 1816 and 1818, and as Calhoun suggests, the drawings in "Moths of Georgia" were probably made a bit later, in the early 1820s. See Calhoun, "Butterfly Drawings in the Hargrett," 125–26. Calhoun points out that the paper for these drawings was made in the United States. Because the American paper industry was in its infancy for much of his career, Abbot tended to use English paper. Not until the 1820s did he use American-made paper with any regularity.
7. For good and bad pencils, see *The Art of Drawing, and Painting in Water-Colours* (1735), 8–9.
8. Trade card of Dorothy Mercier, print seller and stationer at the Golden Ball, London, engraved, ca. 1761, E. 291-1967, Victoria and Albert Museum, London. For abortive vellum, a special kind of vellum (in this case, skin probably from a stillborn or aborted calf), see Rennicks, "The History of Vellum and Parchment" (2022). A spectacularly beautiful copy of Smith and Abbot's *Insects of Georgia* was printed on vellum with several of the illustrations colored by John Harris, engraver, colorist, and artist, son of Moses Harris. It is held in the special collections of University of Georgia's Hargrett Library.
9. For the apothecary shop as a site where medicinal and decorative materials overlap, see, for instance, Paula De Vos, "Apothecaries, Artists, and Artisans" (2015).
10. "To be sold by Robert Bass, Apothecary," *Pennsylvania Gazette*, 27 November 1776.
11. Robert Bass, "Imported this day, the 15th of March . . . from London," *Pennsylvania Gazette*, 2 April 1788.
12. *Republican; and Savannah Evening Ledger*, 20 April 1809.
13. Anson Parson, *Daily Savannah Republican*, 18 December 1828; Hazard and Denslow, *Daily Savannah Republican*, 2 February 1832.
14. *Daily Savannah Republican*, 18 November 1823.
15. *Daily Savannah Republican*, 21 November 1831.
16. *Republican; and Savannah Evening Ledger*, 4 March 1809.
17. For the importance of color in depicting Lepidoptera and the problem of the inconsistency of paints and pigments, see Jovanovic-Kruspel, "The Art of Painting Butterflies" (2021); and Jovanovic-Kruspel and Bruckner, "Ignaz Schiffermüller and the Colors of Butterflies" (2024).
18. *The Art of Drawing* (1757), 64–65.
19. *The Art of Drawing, and Painting in Water-Colours* (1757), title page.
20. Abbot, introduction to the manuscript of *Insects of Georgia*, 6, Smith MSS vol. 39, "Varia Scientific."
21. *The Delights of Flower Painting* (1756), 11.
22. Gum arabic, along with pigments made from minerals, can be bought today online from pigment distributors such as Natural Pigments. It is a fascinating product of ancient origin used today in medicines, candies, and paint. For instance, the hard, shiny surface on the licorice-flavored candy Good and Plenty, once popular in movie theaters, is derived from gum arabic.
23. *The Artists Assistant* (1794), 61. This American recipe book copied, with little modification, the British publication *The Art of Drawing, and Painting in Water-Colours* (1735, 1757, 1768, 1779), for example: "The use of the Coloquintida is only to keep the Flies from spoiling your Work, if it should be exposed" (67).
24. Edwards, *Essays on Natural History* (1770).
25. *The Artists Assistant*, 61.
26. Abbot, introduction to the manuscript for *Insects of Georgia*.
27. *The Art of Drawing* (1757), 82–83. There are many editions of this manual. Abbot used one published in 1757 by George Keith.
28. Pearl ash is potassium carbonate, a product of baking potash, also known as salts of tartar. *The Art of Drawing* (1757), 63.
29. French berries were sold at Lay and Hendrickson, who advertised themselves as "Druggists" in the *Daily Savannah Republican*, 21 November 1831.
30. *Daily Savannah Republican*, 2 February 1832. For more on indigo production in the South, see Feeser, *Red, White, and Black Make Blue* (2013).
31. Abbot's concern about the stability of watercolor paint could not prevent lead-based white paint from turning black over time, as it happened to his own self-portrait. He must have used lead-based paint for highlights on his nose, something that he did not repeat in his entomological portraits. "It is observable that white Lead will turn black, if mixed with Water that comes from Iron or Clay; that is, in the Space of a Month or two, you may perceive those places where it lies thickest tinged with black; and if it be mixed with any other Colour it will soon change, or alter it" (*The Art of Drawing* [1757], 57).

32. For the importance of color to identifying and painting insects, see *Insects and Colors*, edited by V. E. Mandrij and Giulia Simonini.
33. Drury to Harris, 15 March 1770, Drury's Letter-book, fol. 193.
34. Drury, *Illustrations of Natural History* (1773), 2:iv.
35. Albin, *A Natural History of English Insects* (1724).
36. Avinoff, undated transcript, Avinoff Papers, quoted in Wilkinson, "John Abbot's London Years. Part 1," 117.
37. Avinoff, quoted in Wilkinson, 117.
38. Avinoff, quoted in Wilkinson, 117.
39. Avinoff, quoted in Wilkinson, 117.
40. With "this library," Avinoff is referencing his library of entomological treasures (31 October 1945, Avinoff Papers).
41. Abbot drawing, fol. 96, Archives of the Carnegie Museum of Natural History, Pittsburgh, Pa. Of the one hundred drawings, ninety-seven were by Abbot, and of these, Abbot dated ninety-five, and most carry his annotations.
42. Abbot drawing, fol. 37.
43. Abbot drawing, fol. 95.
44. "Neat rows" and "tabular" are Wilkinson's terms ("John Abbot's London Years. Part 1," 115–16).
45. For more information on Emory's volume, see Baker, "John Abbot's Insects of Georgia" (1959). For more information on the butterfly drawings in these volumes, see Calhoun, "Butterfly Drawings in the Hargrett"; Calhoun, "Butterfly Drawings in the Houghton"; Calhoun, "John Abbot's Butterfly Drawings for William Swainson"; and Calhoun, "From Oak Woods and Swamps."
46. Doubleday to Harris, 4 June 1839, Harris Correspondence, quoted in Scudder, *Entomological Correspondence*, 123. Calhoun points out that the Scudder edition of this letter claims that this volume did not belong to Swainson, but the actual letter reads: "They are bound in a small folio volume, & did belong to Swainson" (Doubleday to Harris, 4 June 1839, quoted in Calhoun, "John Abbot's Butterfly Drawings for William Swainson," 15). The number of figures in this volume, according to a penciled notation on the dedication page (probably made by a librarian), are 155 Lepidoptera and 392 Coleoptera, a total of 547 figures.
47. "Note by Samuel H. Scudder" in response to Kirby's response to Scudder's article "John Abbot, the Aurelian" (1888), 232.
48. More of these decorative albums may be in private hands passed down in families, such as the Emory album, which Abbot may have given to a member of the Barnsley family of Adairsville.
49. The paper in the Emory album has the watermark "Butler." This is possibly the watermark of what became the Oakland Mills, founded in 1784 by Butler and Hudson, located in Oakland, Connecticut, which was eventually absorbed into Manchester. See Susan Barlow, "Hockanum River Central to Manchester's History," http://www.manchesterhistory.org/reprints/MHS3_Hockanum.html.
50. J. K. Wadleigh described how she went about identifying insects in these fuzzy drawings from the 1820s and 1830s. Of figure 4.8, she wrote:

> This is one of the trickiest groups to identify.
>
> The locust borer is the most likely species choice out of the two *Megacyllene* it could be, the banded thorax is an indicator for the genus. The other species, *M. caryae* has thinner bands and they usually alternate a yellow and white, the W-shaped pattern usually prominently white. The second one I think actually matches a few species found out west, *M. angulifera* & *M. powersi*—but they're not found in Georgia.
>
> The little one on the left is the most uncertain. Based on the overall body shape and patterning variability, I think *Sternidius* is most likely, but can't say so confidently. Other possible options include: *Sternidius mimeticus/alpha*, *Astylopsis fascipennis/macula*, *Leptostylopsis planidorsus*, and *Leptostylus transversus/asperatus*. As a result, leaving the ID at tribe level is the closest I could get.
>
> The two middle ones certainly look like *Monochamus* pine sawyers again, but I don't know if I can confidently say that they're titillator compared to *M. notatus* and *M. carolinensis* with this drawing—especially since he didn't seem to detail the spines at the tip of the elytra in this one.
>
> The cottonwood borer is one of the only options that has those 3 dots on the thorax and is generally black and white. The patterning for that species is quite variable, though usually has more white on it. They are also pretty rare in Georgia.
>
> The last one, while similar to the locust borer, is likely the amorpha borer. This one is tricky because the thickness of the yellow banding can be extremely variable. But the wider yellow bands and straighter banding pattern towards the end of the elytra points to the amorpha since it's not seen in other species of the genus. They are rare in Georgia, but it is within their historical range. (personal communication, February 2025)

51. For Drury's investment in the visual representation of insects, see Murphy, *Captivity's Collections*, chap. 5.
52. Presumably, the hidden compartment for drawings and notes was to conceal them from customs officials, who would have taxed them. Francillon asked John Leigh Philips, his Manchester friend, to return the box he had used to send Philips some specimens. This box belonged to Abbot, and Francillon needed to return it to him on the next ship. See Francillon to Philips, 2 September 1794, fol. 84, Add. MS 29533.
53. The Drury/Westwood papers held by the Hope Library of Entomology, Museum of Natural History, Oxford University contains a note written in 1952 saying that

Drury loaned Donovan his manuscript catalogs (dated 1784, 1785, 1786) to assist with the writing up of a sales catalog. During this process, Donovan copied from Drury's manuscript catalog the Linnaean names of the insect specimens and their country of origin, but, oddly, he left out the name of the specimen collector. In the 1780s Drury had sent out feelers to naturalists abroad and at home looking for a potential buyer of his collection. A printer's copy of an advertisement for the auction is dated May 1788; this particular copy has large sections crossed out in black ink.

54. See, for instance, the preface to *Insects of India* (1800), where he mentions the "advantages derived from occasionally inspecting the noble collections of Mr. Francillon, Mr. Drury, and some others," among whom was "A. M'Leay" (2).

55. Drury's annotations on fol. 9, MS Typ 426.2.

56. These two diaries are in the manuscript collection of Dru Drury, MSS DRU B 1:1, Entomology Manuscripts, NHM. The second diary's final few pages contain notes written by what seem to be physicians who had conducted a postmortem. They had cut open his bladder, revealing the largest stones they had ever seen in a "human being." There are also notes on his estate, perhaps written by his son, reporting on the June 6, 1805, sale of the insect collection, which garnered £763 13s. 2d., and on "Mr McLay" (Macleay) having bought 735 insects for £27 11s. 6d. on February 10, 1805, which suggests he arranged to buy specimens privately before the auction.

57. The IDs on the back of some of these drawings refer to 1870s and 1880s nomenclature that has since been revised and abandoned. For instance, the drawing of a viceroy butterfly (see figure 5.4) carries on its back the label "*Limenitis misippus*," a name that was bestowed on this species by Strecker in 1878 and that was used by the annotator. It was an update on Cramer's 1775 name *Papilio archippus*. *L. archippus* now seems to be the name of this species, a reference back to Cramer's original designation. Strecker's name was superseded in 1889 by Scudder's *Basilarchia arthechippus*, in 1903 by Dyer's *B. archippus*, and so on. For more information on the evolution of this species' name, see the Limenitis page at the "Lepidoptera and Some Other Life Forms" database, https://ftp.funet.fi/index/Tree_of_life/insecta/lepidoptera/ditrysia/papilionoidea/nymphalidae/limenitidinae/limenitis/.

58. The binding of the Drury volume is quite similar to the volume of Abbot drawings that the Boston Society for Natural History had purchased from the Oemler family. It seems likely that the society rebound or bound these two volumes sometime after 1874.

59. Richard Pulteney, Smith's friend and Linnean Society fellow, donated his natural history collection to the Linnean Society. A few decades later, the society included his collection in an auction of specimens as a space-saving venture, certainly something that Pulteney would not have anticipated when he bequeathed the society his precious collection.

60. For this information, see a bookplate pasted to the inside board of Abbot's 1772 volume. Also pasted to the interior front page are two pages from *Bernard Quaritch's Natural History Catalogue* (n.d.), advertising this volume for twenty pounds. Bernard Quaritch (1819–99) was a London-based bookseller known for dealing in rare books and manuscripts.

61. I suspect these drawings owned by Drury may have entered the British Museum via Donovan as a gift, or the museum may have acquired them from the auction of his collections.

62. A few drawings in the Avinoff collection picture lepidopterans with larvae, pupae, and host plants. The 1772 volume displays insect bodies in a tabular arrangement only. Wilkinson notes that between 1766 and 1773, when Abbot was in London, he drew "only seven of the nearly one hundred and fifty sheets of illustrations" in a style reminiscent of Eleazar Albin's *Natural History of English Insects* (1720, 1724), a style that included what was thought to be host plants along with the Lepidoptera in various stages of metamorphosis. See "John Abbot's London Years. Part 1," 115.

63. For more on Merian's illustrations featuring caterpillars, see Etheridge, *The Flowering of Ecology* (2021); and Jovanovic-Kruspel, "The Art of Painting Butterflies." Jovanovic-Kruspel also addresses the shift in drawings from depicting insects as if they were specimens in a drawer to depicting them as living creatures.

64. Haworth, "Science of Entomology in Great Britain" (1812), quoted in Wilkinson, "John Abbot's London Years. Part 3," 229, which gives the publication date as 1807.

65. In the preface Wilkes acknowledged that he knew he should depict the host plants of the caterpillars he featured, but he often chose an exotic plant to portray because it was more interesting and attractive than the usual nettles or oaks that many of his subjects ate.

66. Swainson, *Taxidermy*, 99–100.

67. Oemler to Harris, 14 March 1851, Harris Correspondence.

68. Oemler's album of 193 drawings was purchased in 1874 by the Boston Society of Natural History from Arminius Oemler, son of Abbot's friend Augustus Oemler. The drawings are bound together with Abbot's original "Notes to the Drawings of Insects," along with a copy of Abbot's notes, which have been reordered, renumbered, and entitled "Notes on the drawings of Insects, by Mr. John Abbot, in Dr. Oemler's possession." The handwriting on the drawings, I assume, is the elder Oemler's, while the copyist of Abbot's notes is unknown and in a completely different hand, perhaps belonging to a curator or librarian at the Boston Society. The album is now in Harvard's Houghton Library, MS Typ 426.3.

69. Abbot to Swainson, 7 June 1819, Swainson Correspondence.
70. Calhoun suggests that on the few occasions when Abbot misidentified a plant as a species' host plant, he may have purposefully chosen to draw a more attractive plant. See Calhoun, "John Abbot's Butterfly Drawings for William Swainson."
71. Scudder hypothesized that Abbot raised over sixty species of butterflies ("John Abbot, the Aurelian," 153). The number of moths he raised is, of course, much higher.
72. Abbot to Escher, 18 April 1813, Savannah, Abbot Letters to Escher. Abbot mentions in his correspondence with Escher that several of his letters seem to have gone missing, and as a consequence, it is difficult to reconstruct with accuracy when and how many drawings Abbot sent him. In September 1813 he mentions that he has on hand 150 drawings, which Oemler had hoped the Savannah Library could afford to buy: "I have some Time past, sold 80 such Dss. to the S. Library, all not figured in Smith. These 150 contain about 100 more Species not in Smith[.] One reason for not answering your Letter sooner, Mr Oemler is still desirous the Library should have them, but the funds is quite deficient owing to the Calamitous War which affects everything & which is not likely to be soon remidied" (Abbot to Escher, 3 September 1813, Abbot Letters to Escher). The library managers may have failed to close the deal with Abbot in a timely manner due to the economic uncertainty caused by the War of 1812, and Escher could have acquired some of these 150 drawings.
73. Escher's collection of Abbot's drawings has been scanned and is available for viewing online: Abbot, John: Dessins Originaux de Plantes et Insectes de Géorgie [Georgia], [zwischen 1776 und 1840], Zentralbibliothek Zürich, NB 43, https://doi.org/10.7891/e-manuscripta-96325 / Public Domain Mark. For more information on the butterfly drawings in Escher's collection, see Calhoun, "From Oak Woods and Swamps."
74. These drawings have been reproduced in Neri, Nummedal, and Calhoun, *John Abbot & William Swainson*. Abbot sent Swainson 104 drawings, but one seems to have been lost. Abbot to Swainson, 1 May 1818, Swainson Correspondence.
75. Swainson to Abbot, 25 October 1817, Bahia Brasil, Swainson Correspondence.
76. Abbot provided a footnote to this letter: "+ Except that I shall draw among them some of the other Genera of Insects" (Abbot to Swainson, 20 December 1816, Savannah, Swainson Correspondence). This explains something I've always thought was a bit odd: The Swainson and Hargrett flora et entomologia drawings include a few nonlepidopterous insects; both begin with a few beetles and a cricket and end with a dirt dauber wasp. After looking at these drawings for years, I realized that all the insects depicted in these particular drawings were raised by Abbot. While it is common to find in breeding journals dates of emergence from the chrysalis for moths and butterflies, it strikes me as unusual to have a record of a beetle's transformation from grub to bug and even more unusual to know when a wasp emerged from its larval state in its mud-built cell.
77. Swainson writes of his "intention of forming two additional volumes to those edited by Dr. Smith: but the design is now abandoned" (*Taxidermy*, 100).
78. His espousal of the quinarian taxonomic system, which grouped plants and animals in sets of five based on notions of affinity, may have cost him the respect of his peers.
79. For more information on William Swainson, see Nora F. McMillan's collection of documents she gathered and notes she wrote in preparation for a biography of William Swainson. McMillan presented these documents (four boxes) to the Natural History Museum's library. They are referred to as "Swainsoniana."
80. This number does not take into account that there may be other such albums or individual drawings in private hands. For instance, Deen Day Sanders owns twelve life history drawings (Calhoun, "From Oak Woods and Swamps"). I have personally seen Swainson's life history drawings held in New Zealand's National Library; Francillon's and Sir Robert Johnson Eden's, Fifth Baronet of West Auckland, in London's Natural History Museum; Drury's and Oemler's in Harvard's Houghton Library; McDougal's in the University of Georgia's Hargrett Library; and the originals at Johns Hopkins University. See Calhoun, "John Abbot's 'Lost' Drawings," for a description of Boisduval's missing drawings. Calhoun reports in "From Oak Woods and Swamps" that these drawings were never missing but had been loaned to Boisduval and returned to their owner, Escher, and are now in the Zentralbibliothek Zürich. Missing are the drawings that the Savannah Library Society and the Charleston Library owned. According to Rogers-Price, the Savannah Library Society purchased eighty drawings in 1813. In an 1817 letter to Stephen Elliot, Abbot asked "if the Charlestown Library wou'd take them, as I have now 104 finished for a 2 Vol" (Abbot to Elliot, 3 August 1817, Stephen Elliot Papers, gra00020, Archives of the Gray Herbarium, Arnold Arboretum Library, Harvard University, https://nrs.lib.harvard.edu/urn-3:fmus.gra:32691737).
81. For use of templates, see Gilbert, *John Abbot*.
82. The plants in Swainson's and Hargrett's drawings of the giant swallowtail are Hercules-club (*Zanthoxylum clava-herculis* L.).
83. Abbot to Elliot, 3 August 1817, Elliot Papers.
84. Neri, "Introduction: Science and Sentiment," 1, 21.
85. Escher's collection of 201 drawings shares images with those in Francillon's volumes 16 and 17. For instance, Francillon's drawing of the great swallowtail butterfly has an orange tree in it, as does Escher's. See chapter 5 for more on Escher's collection of Abbot's life drawings.
86. See Calhoun, "From Oak Woods and Swamps," for the details concerning the dating of Abbot's drawings.
87. For more on Swainson's motives, see Neri, "Introduction: Science and Sentiment."
88. See Rogers-Price's catalog for a mention of private col-

lections of Abbot drawings, such as a "volume of thirty-eight watercolors" found in the notes accompanying figure 11 (*John Abbot in Georgia*, 61).

Chapter 5. Francillon and LeConte

1. Abbot to Oemler, 26 September 1833, Bulloch County, Harris Correspondence, https://nrs.lib.harvard.edu/urn-3:fmus.mcz:29995360.
2. Abbot's use of the term "quarto" refers to the size of a sheet of paper, a term derived from early printing methods that involved large pieces of paper printed and folded into leaves. A quarto is approximately 24 × 30 centimeters or 9.5 × 12 inches.
3. For a discussion of terminology, including phrases such as "a man of science," used to describe those who practiced science before the word "scientist" became common in the latter half of the nineteenth century, see Wale, *Making Entomologists* (2022).
4. Francillon was a British subject of Huguenot heritage, and LeConte was an American citizen also of Huguenot heritage.
5. Francillon to Philips, 10 August 1793, Add. MS 29533, fol. 81v.
6. Francillon to Swainson, 15 April 1816, Swainson Correspondence.
7. Francillon to Philips, 22 August 1794, Add. MS 29533, fol. 83.
8. Francillon to Philips, 2 September 1794, Add. MS 29533, fol. 84.
9. Francillon to Philips, 22 October 1794, Add. MS 29533, fols. 85, 86.
10. Francillon to Philips, 1 November 1805, Add. MS 29533, fol. 96.
11. Francillon's claim that there were only two sets of these drawings might have been true in 1806, but by 1813 Abbot was promising Escher similar drawings and Swainson likewise in 1818.
12. Francillon to Philips, 13 January 1806, Add. MS 29533, fol. 98–98v.
13. These and other spider drawings are discussed in the concluding chapter.
14. Francillon sent Philips a box of insects Abbot had collected: "I think [they] will afford you pleasure when you see them" (Francillon to Philips, 17 November 1790, Add. MS 29533, fol. 72). In this letter Francillon lays out what Abbot's bill amounted to—243 insects at sixpence each—plus duty of 30 percent that he paid. He does not ask for a finder's fee or a percentage of the cost.
15. Francillon told Philips about meeting Bracy Clarke, visiting London from Worcester, and how "during his stay in London, I gave him every Assistance in my power and introduced him to Mr Marsham, Doctor Smith, and every collection worth his notice" (Francillon to Philips, 11 November 1791, Add. MS 29533, fol. 76).
16. Francillon to Philips, 19 August 1786, Add. MS 29533, fol. 66.
17. Francillon to Philips, 28 August 1786, Add. MS 29533, fol. 67v.
18. Francillon to Philips, 10 August 1793, Add. MS 29533, fol. 81v.
19. Francillon to Philips, 10 August 1793.
20. Francillon to Philips, 29 August 1787, Add. MS 29533, fols. 70–71. Fabricius visited London several times examining specimens in British insect collections, including Drury's, the Duchess of Portland's, and Francillon's. For more on Fabricius and his connections with British entomology, see Hancock, "The Shaping Role of Johann Christian Fabricius (1745–1808)" (2015), 151–63.
21. This list is from W. F. Kirby's reply to Scudder's article ("John Abbot, the Aurelian") in which Scudder had muddled a number of details that Kirby, as a curator at the British Museum, felt he must correct (Kirby, "John Abbot, the Aurelian," 231–32). The number of drawings is taken from an 1883 Natural History Museum memorandum quoted in Gilbert, *John Abbot*, 108.
22. Yale, *Sociable Knowledge*, 213.
23. Data on Francillon's collection are from the NHM catalog: 2,843 watercolor drawings on Whatman paper, seventeen volumes from 1792 to 1804, purchased by the British Museum after Francillon's death in 1818 for £300 (£26,000 in today's money, but worth much more—a copy of *Insects of Georgia* is worth that). The arthropod drawings were transferred to South Kensington in 1881.
24. Abbot to Swainson, 20 December 1816, Savannah, Swainson Correspondence. John Eatton LeConte did not own plantations in Georgia or anywhere else. His older brother Louis inherited a rice plantation in Georgia. About the various spellings of LeConte, one reason why John Eatton LeConte's publications in France carry the name Leconte instead of LeConte is that in France the capitalization of the letter *C* implied an aristocratic lineage, and it was illegal to publish with a name that made a fraudulent claim on such a rank. For more explanation, see Macknight, *Aristocratic Families in Republican France, 1870–1940* (2012).
25. Louis LeConte, father of scientists John and Joseph LeConte, was an amateur naturalist and gifted garden designer. The LeConte family was originally from New Jersey, where both Louis and John Eatton were born and raised. John Eatton resided in New York City until 1852, when he moved to Philadelphia. During the Civil War, John Eatton's son, John Lawrence LeConte (1825–83), joined the U.S. Army Medical Corps as a surgeon, while his cousins, John (1818–91) and Joseph (1823–1901), joined the Confederate Army as munitions experts. For more information about John Lawrence LeConte, see Samuel H. Scudder, "Dr. John Lawrence LeConte, a Memoir Read to the National Academy of Science, April 17, 1884" (1884).
26. Abbot to Thaddeus W. Harris, 30 August 1835, Bulloch County, Georgia, in Dow, "John Abbot, of Georgia," 71.
27. Abbot to Harris. LeConte was paying Abbot 6.25 cents per drawing.

28. Oemler to Harris, 27 April 1834, Savannah, in Dow, "John Abbot, of Georgia," 70.
29. Oemler to Harris, 1851, Mayr Library, quoted in Calhoun, "John Abbot's 'Lost' Drawings," 216.
30. In addition to the many copies held in libraries of land grant universities in states such as Iowa, Kansas, Michigan, Missouri, and Minnesota, the National Agricultural Library of the U.S. Department of Agriculture holds three copies. For the publication dates of *Histoire générale et iconographie des lépidoptères et des chenilles de l'Amérique septentrionale*, which are various, see Cyril F. dos Passos, "The Dates of Publication" (1962); and C. F. Cowan, "Boisduval and Le Conte" (1969). The dates I've cited are those on the University of Massachusetts's copy of the book.
31. Of these sixty-two plates, five or six may not have been drawn by Abbot. Calhoun argues that plates 3, 14, 15, 17, and 23 featuring non–North American Lepidoptera could not have been drawn by Abbot. See Calhoun, "*Histoire générale*" (2004).
32. For an analysis of how Boisduval's artists and engravers rearranged figures from Abbot's drawings for the illustrations in his and LeConte's publication, see Rogers-Price, *John Abbot in Georgia*, 37–38, 131–39. For details about the complex process of preparing these illustrations for publication, see Calhoun, "*Histoire générale*."
33. Abbot explained to Escher that LeConte requested drawings of butterflies and moths and their caterpillars and chrysalides but "without the plant" (Abbot to Escher, 1 September 1813, Abbot Letters to Escher).
34. These numbers are based on the University of Massachusetts's copy of Boisduval and LeConte's book, from which plate 6 is missing. Calhoun compared the original drawings and the book's plates. He argues that "the wrong artist was credited on seven plates" and that fifty-seven plates can be attributed entirely to Abbot, four plates combine Abbot's figures with Blanchard's, ten plates were drawn by Blanchard, and seven were drawn by Dumenil. See Calhoun, "*Histoire générale*," 154.
35. Boisduval wrote in April 1829 to LeConte, who had returned to New York City, about the progress he was making on their book: "Our book . . . is well on its way with the first volume due out this month . . . and the others to follow" (my translation). However, he did complain about the inexact process of producing colored illustrations of lepidopterans. There was dissatisfaction with the illustrations based on the drawings sent to him by M. Escher; they were difficult to color properly—"faire le coloris d'apres la nature"—because he did not know what the living creatures looked like. See Boisduval to LeConte, April 1829, Paris, John L. LeConte Papers, 1812–97, MSS.B.L493, Library and Museum, APS. What Boisduval is getting at is the problem European collectors faced when trying to identify species that they knew only as collected specimens. They were dependent on Abbot's drawings of larvae, pupae, and in some cases host plants to identify their cabinet specimens of North American Lepidoptera. See Tobin, "Color, Taxonomy, and Exotic Specimens" (2024).
36. Boisduval and Guenée, *Histoire naturelle des insectes* (1836–57). The first volume is accompanied by an atlas (1858) authored by Boisduval and Guenée, with twenty-four plates and several figures on each plate. Plate 2 has fifteen figures; all except for figure 5 were copied from Abbot's drawings: "Toutes ces figures sont copiées sur des dessins originaux d'Abbot, à l'exception de la figure 5" (3).
37. For a detailed analysis of the twenty-three figures that Scudder copied and later used in his own publications, see Calhoun, "*Histoire générale*."
38. Scudder, "John Abbot, the Aurelian," 153.
39. See Calhoun's detailed description of the alterations made by Boisduval's artists and engravers to these drawings in his article on Abbot's butterfly specimens and drawings, "From Oak Woods and Swamps," app. 1.
40. Calhoun, "History of the Lepidoptera Illustrations in the Thomas Cooper Library, University of South Carolina" (2004), 2.
41. Calhoun suggested in 2004 that the missing sets of flora et entomologia drawings were included in this transfer in the 1920s and the eventual Sotheby's sale. The Sotheby's sale catalog does not support this theory, as it listed two volumes of moth drawings (thought to be lost) and one volume of butterflies and moths with their caterpillars (now at the University of South Carolina). See Cowan, "Boisduval and Le Conte," for this description. Calhoun's further research led to the discovery of these missing drawings. See his article "From Oak Woods and Swamps," which describes the missing sets of flora et entomologia drawings that had belonged to Escher and were never a part of the Sotheby sale.
42. To see these drawings, explore the very impressive website, John Abbot Watercolors, located within the University of South Carolina's Digital Collections: https://digital.tcl.sc.edu/digital/collection/abbot/search/page/1.
43. For the publication of these butterfly drawings, see Calhoun, "*Histoire générale*."
44. Written on one of the opening pages is "Dessins originaux de John Leconte pour son l'ouvrage sur les lépidoptères de l'Amérique du Nord, en collaboration avec le Docteur Boisduval." The bookseller who sold the University of Connecticut these volumes is Howard S. Mott, whose name is written in pencil on the back of volume 1.
45. Some of the drawings are overtrimmed, and some may have had LeConte's initials sliced off in the process. This is the case with volume 1, fol. 82, and could be also with volume 1, fol. 88. The number of LeConte drawings is therefore not exact.
46. Another group of moth drawings that may possibly be among those LeConte brought to France are two volumes held in the Entomology Library of London's Natural History Museum. Bound in green leather with the title

"Water Colour Drawings of the Lepidoptera of Georgia" embossed on their spines, these two volumes contain drawings of single moths on small pieces of paper pasted onto larger folio-sized paper, much like those in the University of Connecticut Special Collections. Pasted inside both volumes is a manuscript note: "Drawings by Abbot only." This note might work to differentiate these two volumes from those that contain a mixture of LeConte and Abbot drawings held by the University of Connecticut. More research needs to be done to work out the provenance of these two different sets of moth drawings. See the Natural History Museum's library description of the two moth volumes, "John Abbott (1751–1840) collection of 248 watercolours for the published work Lepidoptera of Georgia [1790s?]," https://nhm.primo.exlibrisgroup.com/discovery/fulldisplay?docid=alma9932530902081&context=L&vid=44NHM_INST:44NHM_V1&lang=en&search_scope=MyInst_and_CI&adaptor=Local Search Engine&tab=Everything&query=any,contains,Abbot John&sortby=date_a&mode=basic.

47. These small drawings of moths in figure 5.12 are difficult to identify, as J. K. Wadleigh attested:

> This one was quite the investigation! I've concluded that the only reasonable ID is of the genus *Proserpinus*. Initially I looked at quite a few other GA genera that weren't coming close enough to the drawing. *P. gaurae* is the only species within the genus that has a historical range into GA (and most of the SE in general), but it has almost no modern records there so it was harder to track down!
>
> Fairly confident in the Delaware skipper ID. It's the only species of that genus in GA, is pretty common, and has a pretty unique patterning. The faint dorsal wing vein patterning is specific to this species. Any similar species for that region of the US have much darker patterning pretty consistently and/or some underwing patterning as well. It's very likely a male with the reduced patterning, too. (personal communication, February 2025)

48. I am in agreement with Calhoun, who confirms Rehn's supposition that many Abbot drawings are to be found in the library at the American Philosophical Society under the rubric of LeConte's entomological drawings. See Calhoun, "John Abbot's 'Lost' Drawings." For a description of the APS's collection of LeConte drawings, see Rehn, "The John Eatton Leconte Collection of Paintings of Insects, Arachnids, and Myriopods" (1954).
49. In 1841, one year after Abbot's death, Spencer Fullerton Baird (1823–87), a distant cousin of the LeContes and the Smithsonian's first curator, spent a day "looking over major Leconte's Entomological drawings, of which he has about 9000 sheets, a species to each sheet; they are most beautifully executed" (Dall, *Spencer Fullerton Baird, a Biography* [1915], 58). This figure of nine thousands suggests that there are many more drawings that are still unaccounted for that LeConte had collected and perhaps had drawn himself.
50. Rehn's count is 4,043; Calhoun's is 3,700. The number figures for each volume were copied from the annotated front matter of each volume. See the APS's description of these drawings: "Entomological Drawings, n.d. Mss.595.7.L493," https://as.amphilsoc.org/.
51. Abbot to Escher, 31 December 1813, Abbot Letters to Escher.
52. Scudder, "Dr. John Lawrence LeConte," iv. Rehn suggests that John Lawrence LeConte was the artist who produced seventy-one beautiful graphite beetle drawings: "It is very probable that these figures were prepared by John Lawrence Leconte for inclusion in a report on the Coleoptera taken by the Mexican Boundary Survey" ("The John Eatton Leconte Collection," 447). Abbot, of course, does not seem to have traveled much beyond Georgia, even though he mentioned his desire to go to Florida and Tennessee.
53. The annotations are not limited to the LeContes, with many different hands offering their best guess at the proper Linnaean name of these insects. The provenance of these drawings can be traced in the dealer advertisements that the APS has preserved, perhaps once pasted onto the boards of the portfolios. See also the APS's Archive Ascension Records, vol. 10, March 1952–June 1954.
54. T. W. Harris to J. E. LeConte, 17 December 1829, LeConte Papers. Emphasis is Harris's.
55. Abbot wrote Escher about LeConte's commissions: "First [LeConte] engaged me to make a General Collection of & draw all the Coleoptera of Georgia for him, Then all the Lepidoptera, The Papilio [butterflies], the Cat^r. Chry^s. & Fly, without the plant, The Phalena [moths], Male & female when different without the plant, & all the Aranea [spiders]" (Abbot to Escher, 1 September 1813, Abbot Letters to Escher).
56. "Inferior copyist" is Scudder's phrase, which he used to reject the idea that Abbot was the artist of the volume of drawings that Doubleday acquired in a London bookshop and gave to Harris. Doubleday states in a letter to Harris: "I found at a Booksellers 84 drawings by Abbot, containing 150 figures of Georgian Coleoptera & about 350 Lepidoptera. They are bound in a small folio volume, & did belong to Swainson" (Doubleday to Harris, 4 June 1839, in Calhoun, "Inferior Copyist," 228). Calhoun notes that when Scudder published Harris's entomological correspondence, he revised Doubeday's letter to read "did not belong to Swainson."
57. Harris to LeConte, 17 December 1829, LeConte Papers.
58. See LeConte and Horn, *Classification of the Coleoptera of North America* (1883).
59. Gray, "Some North American Botanists" (1883), 199. LeConte senior complained to Swainson of his inability to write due to arthritic pain in his hands. See LeConte to

Swainson, 25 August 1827, New York, Swainson Correspondence.

60. Gray, "Some North American Botanists," 198.
61. Henkels, "Catalogue of the Valuable Entomological Library of the late John L. Le Conte, M.D. of Philadelphia" (1884). The American Entomological Society (J. L. LeConte was the founder) is named as the depositor. Does this mean that the American Entomological Society was divesting itself of what may have been a LeConte bequest? Other LeConte bequests, specimens mostly, were welcomed by the Academy of Natural Sciences. According to Gray, John Eatton LeConte's "extensive and valuable herbarium . . . was presented to the Academy of Natural Sciences of Philadelphia in 1852, and was followed shortly after his death by a large collection of fresh water mollusca of the United States, containing many original specimens of species first observed by him" (198).
62. For a nuanced account of the tensions generated over who possessed expertise within entomological communities in nineteenth-century Britain, see Wale, *Making Entomologists*.
63. Referring to J. L. LeConte's labors as a coleopterist, Scudder asserted: "Nearly half of our Coleoptera have been described for the first time by him (he actually described or at least named 4,739 nominal species). . . . That LeConte was the greatest entomologist this country has yet produced is unquestionable" ("Dr. John Lawrence LeConte," xvi–xvii).
64. J. L. LeConte, introduction to LeConte and Horn, *Classification of the Coleoptera*, xxviii. This treatise contains no illustrations except for five diagrams of beetle body parts (serrate and clavate antennae, mandibles, thoraxes, and legs) designed to "make a brief exposition of the external anatomy of Coleopterous insects, before attempting to define the numerous families which compose the order" (xi).
65. Henkels, "Catalogue of the Valuable Entomological Library," 17. The total number of entomological drawings for sale at the auction comes to 4,676, while the total number of entomological drawings at the APS is 3,695. These numbers suggest that nearly a thousand drawings have gone missing since the auction.
66. The De Renne family of Savannah has for several generations supported the study of Georgia's history as writers, publishers, and collectors of books, maps, documents, and manuscript materials pertaining to Georgia's past. For more information on the De Renne family's collection of books and manuscripts, see Bragg, *De Renne* (1999).
67. Hargrett acquired LeConte's frog and turtle drawings in 1938 from the De Renne family, who had purchased them in 1905 from S. N. Rhoads, a natural history dealer in Philadelphia. In 1907 Rhoads sold J. E. LeConte's other drawings, which included Abbot's, to the Missouri Botanical Garden for $225. For details about the provenance of these LeConte volumes after the 1884 auction, see Calhoun, "*Histoire générale*."
68. Rehn, "The John Eatton Leconte Collection," 444.
69. See *Catalogue of a Fine Collection of Rare Books and Drawings* (1963).

Chapter 6. "Beauty and Accuracy"

1. Max Barclay, personal communication, December 2014; Joe McHugh, email message, 9 February 2015.
2. 31 October 1945, Avinoff Papers.
3. Daston, "The Empire of Observation" (2011), 99.
4. Daston, 99.
5. Smith, preface to Smith and Abbot, *Insects of Georgia*, ii.
6. The anonymous author (most likely Edward Donovan) continues: "We cannot, however, forbear congratulating the dilettante and the student on the pleasure and information they are about to receive from a sedulous perusal and judicious contemplation of such an assemblage of natural curiosities; and we return our thanks to the publisher, equally for the spirit with which he rescued so valuable a collection from obscurity, and the perseverance and taste with which he superintended the execution of the whole" (anonymous review of *Insects of Georgia*, *Analytical Review* 27 [1798]: 5–6).
7. William Jones to Smith, 9 September 1797, Smith Correspondence, Linnean Society. Smith drew upon Jones's extensive knowledge of butterflies to help him with his taxonomic task. Jones's *Icones*, watercolor drawings of the domestic and exotic butterflies in his large collection and in other collections, including Drury's, Banks's, and Francillon's, are housed at the Oxford University Museum of Natural History. It is possible that some of the North American butterflies in Drury's and Francillon's collections that were painted by Jones had been collected by Abbot. By the mid-1780s, Abbot was supplying Londoners with his specimens. Jones's drawings have been digitized and are available on the museum's website, and they have been reproduced in *Iconotypes* (2021).
8. Swainson, *Taxidermy*, 100; and Swainson, *A Preliminary Discourse on the Study of Natural History* (1834), 68.
9. Haworth, "Review of the Rise and Progress of the Science of Entomology in Great Britain" (1807), quoted in Wilkinson, "John Abbot's London Years. Part 3," 229.
10. The word "entomology" was not used with much frequency in Britain until after 1820. Kirby and Spence's *Introduction to Entomology* (1815–26) is considered a foundational text for what became the field of entomology.
11. "Learned" and "practical" botanists are Smith's terms. See "A Review of the Modern State of Botany," in Smith, *Memoir and Correspondence of the Late Sir James Edward Smith* (1832), 2:441–591, 447–49. Smith held a hierarchical view of natural knowledge practices, with the learned or "philosophical botanist" (591) held in the highest regard, followed by "the half-learned or half-experienced botanist" and "the superficial gatherer of simples or flowers"

(447). For more on these divisions within those who studied botany, see Endersby, *Imperial Nature*.

12. For instance, Ann Shelby Blum addresses the question of why her book, *Picturing Nature* (1993), a history of illustrated natural histories, does not include "the work of naturalists of the early period, such as John Abbot of Georgia, an Englishman, or Nicholas Hentz, a Frenchman, though I did include the work of Alexander Wilson, a Scot, and Charles Alexandre Lesueur, another Frenchman." She explains that her interest is in the publication of scientific illustrations in the United States, and she excluded Abbot because "in the colonial tradition, Abbot sent his commissioned drawings back to England for publication" (xxix). Her focus on American publishing does not, however, preclude a lengthy discussion of Audubon's double-elephant folio *Birds of America*, which was published in Great Britain.
13. Smith, preface to Smith and Abbot, *Insects of Georgia*, ii.
14. Swainson's statement that Abbot raised the "whole of Lepidoptera" is a bit of an overstatement. There are approximately 14,300 lepidopteran species in North America. See "Lepidoptera of North America Network," IDigBio, https://www.idigbio.org/wiki/index.php/Lepidoptera_of_North_America_Network.
15. James Duncan notes that in almost every illustration "both sexes are figured, along with the larva, and the plant it frequents" (introduction to *Natural History of Exotic Moths*, 71).
16. Smith and Abbot, *Insects of Georgia*, 1:3.
17. Smith and Abbot, 2:141.
18. Smith and Abbot, 2:105.
19. Donald W. Hall, University of Florida, Department of Entomology and Nematology, UF/IFAS, Publication Number EENY-464, "Featured Creatures," November 2012, updated February 2019. See a photo of this hairy caterpillar at the UF/IFAS "Featured Creatures" page, https://entnemdept.ufl.edu/creatures/misc/moths/puss.htm.
20. Smith and Abbot, *Insects of Georgia*, 2:105.
21. Smith and Abbot, 1:7.
22. Duncan is quite critical of the failure of Drury's *Exotic Insects* to depict or describe "the different states of the species, in which the original work was wholly defective, not a single lepidopterous larva being either figured or described" (*Natural History of Exotic Moths*, 66).
23. Duncan, 62–63.
24. Drury to Abbot, 10 April 1774, London, Drury's Letter-book.
25. Smith, preface, ii.
26. Smith, iii–iv.
27. Using caterpillars to help identify a species of moth or butterfly can be tricky, since the color of a caterpillar can vary across instars (the various stages of molting) and because of the food the caterpillar consumes. Benjamin Wilkes, author of *English Moths and Butterflies* (ca. 1744, 1773), argued that Lepidoptera should be classified according to the caterpillar's morphology and not the morphology of "adult" butterflies and moths (imago).
28. For a thorough discussion of the accuracy of Abbot's choice in host plants, see Calhoun, "A Glimpse into a 'Flora et Entomologia.'"
29. Smith and Abbot, *Insects of Georgia*, TAB. LIII, 2:105. The current common name of the adult moth is southern flannel moth, and the caterpillar is called the puss caterpillar. Among the hairs on the caterpillar's body are some venomous spines, which can cause a painful reaction. Donald W. Hall of the University of Florida confirms Abbot's description of "lid (door) on the cocoon." For more information, see UF/IFAS "Featured Creatures" page, https://entnemdept.ufl.edu/creatures/misc/moths/puss.htm.
30. Wilkinson, "John Abbot's London Years. Part 3," 223.
31. Smith, preface, ii.
32. Dow, "John Abbot, of Georgia," 66.
33. Abbot's manuscript "A Natural History of the North American Insects," Smith MSS vol. 39, "Varia Scientific."
34. "Smith received rough notes and coloured drawings, probably through the agency of John Francillon" (Wilkinson, "Smith and Abbot," 213).
35. Calhoun, "From Oak Woods and Swamps," and Wilkinson, "Smith and Abbot," speculate that Edwards acquired the drawings from Francillon and chose Smith to act as editor and taxonomist for the volume. In "From Oak Woods and Swamps" Calhoun counters his earlier interpretation of Smith and Abbot's collaboration, but I remain persuaded by Calhoun's earlier assertion that Abbot was "apparently aware that the drawings were going to Smith" (Calhoun, "A Glimpse into a 'Flora et Entomologia,'" 3).
36. John Francillon, *Description of a Rare Scarabaeus from Potosi, in South America* (1795). For more about Francillon, see Cowan, "John Francillon, F.L.S., a Few Facts" (1986).
37. See Calhoun, "A Glimpse into a 'Flora et Entomologia,'" on *Insects of Georgia* for examples of Smith's editing.
38. Smith and Abbot, *Insects of Georgia*, 1:19. Here is Abbot's original note: "The catterpillar was taken the begining of June, feeding on the Wild Gooseberry, It also eats Wild Cherry, and Willow. The 8th June it Spun up by the tail, and the 9th changed into Chrysalis, The Butterfly was bred the 18th. This Butterfly likewise breeds early in the Spring, having taken them 19th April. it is also in Virginia, but neither the Butterfly or Catterpiller is very common. Note the figure of the Catterpiller might be a size larger" (no. 53, "Black and Blue Admirable Butterfly," in "A Natural History of North American Insects," Smith MSS vol. 39, "Varia Scientific").
39. Smith and Abbot, *Insects of Georgia*, 1:19. Smith references William Jones's article "A new Arrangement of Papilios" (1794).
40. The French translation was provided by "Romet," who according to Calhoun also translated the 1794 edition of Moses Harris's *The Aurelian*. See Calhoun, "A Glimpse into a 'Flora et Entomologia,'" 1.

41. Smith and Abbot, *Insects of Georgia*, TAB. CIV, 2:207, TAB. XIII, 1:25.
42. Smith and Abbot, TAB. XIII, 1:25. For Jones's drawings, see *Iconotypes*.
43. Smith and Abbot, *Insects of Georgia*, TAB. LXII, 2:123.
44. Smith and Abbot, TAB. LXII, 2:123.
45. Smith and Abbot, TAB. XXXIV, 1:67.
46. Smith and Abbot, 1:67. The International Commission on of Zoological Nomenclature was formally established in the 1890s; however, the principle of priority operated much earlier within zoological circles that adhered to the principles of Linnaean taxonomy. Smith subscribed to the principle of priority (usually). In this passage Smith is referencing Linnaeus's *Systema Naturae*, 10th ed. (1758).
47. Smith's method of naming is not unlike Wilkes's pre-Linnaean rationale. See Wilkes's preface to *English Moths and Butterflies* explaining his naming of Lepidoptera:

    > To make the Name, if possible, descriptive of the Thing: Which Rule has been followed as strictly as might be in naming the Moths and butterflies contained in this Work: that is to say, the Shape, the Colour, the Marking, the Food, or the Place where found is what they are named from. But in doing this, where the Author found them named to his Hands in this descriptive Manner, he has gladly retained the Name they had already, and only presumes to bestow Names himself on such as either had none at all, or which he thought not sufficiently denotative of the particular Flies intended to be known by them. (n.p.)

48. Smith and Abbot, *Insects of Georgia*, 1:67.
49. Smith and Abbot, 1:69. Smith cites Cramer, *De Uitlandsche Kapellen*, vol. 107, fol. CI.
50. Smith, preface, iv.
51. Smith and Abbot, *Insects of Georgia*, TAB. LXXVIII, 2:175. Smith is referencing *Insecten-Belustigung* (1746–61) by August Johann Rösel von Rosenhof (1705–59).
52. Smith and Abbot, TAB. XV, 1:31 and TAB. XXI, 1:41.
53. Smith and Abbot, TAB. XXVII, 1:53.
54. Daston and Galison, *Objectivity*, 82–88.
55. Drury, preface to *Illustrations of Natural History*, 1:iii–xvi. Fabricius examined Drury's insect collection and helped to classify many of Drury's specimens. For more on Fabricius and his connections with British entomologists and collectors, see Hancock, "The Shaping Role."
56. Drury, preface, 1:xii.
57. Drury, 1:xiv–v.
58. Drury, 1:xv.
59. Drury to Harris, 5 April 1770, fol. 196, Drury's Letter-book.
60. Jardine, "Memoir of Dru Drury," 17.
61. Smith and Abbot, "Wild-Grape Hawkmoth," in *Insects of Georgia*, TAB. XXVIII, 1:55.
62. For the history of the various versions of what became the Royal Entomological Society in the 1830s, see Salmon, *The Aurelian Legacy* (2000).
63. Donovan's attack, published in *Dr. Rees's New Cyclopaedia* (vol. 13, pt. 1), is quoted in Burrell, "A Short Letter to E. Donovan, F.L.S." (n.d.), 6–7. This letter is dated March 1810.
64. Burrell, 12.
65. Burrell, 13.
66. Burrell, 15–17.
67. This manuscript is dated 1772; it is in Harvard's Houghton Library.
68. Wilkinson, "John Abbot's London Years. Part 3," 223.
69. For the class dynamics at work in nineteenth-century natural history communities, see Secord, "Corresponding Interests" (1994).
70. Burrell, "A Short Letter," 16.
71. Burrell, 14, 15, 16, 17.
72. Haworth, "Review of the Rise," vi. Thomas Marsham published the first volume only (in Latin): *Entomologia Britannica* (1802).
73. Haworth, "Review of the Rise," vi.
74. For discussions of some of these tensions between field naturalists and taxonomists in the early twentieth century, see Hancock, "John Russell Malloch: Amateur Naturalist to Professional Taxonomist," in MacGregor, *Naturalists in the Field*, 730–58.
75. Burrell, "A Short Letter," 16.
76. For a discussion of how nineteenth-century scientific communities were formed in ways that complicate the distinction between amateur and professional scientists, see Belknap, "Illustrating Natural History" (2018); and Wale, "Editing Entomology" (2019). In the United States and Britain, those once considered amateur naturalists are now called "citizen scientists," a term that graciously extends some legitimacy over the work of counting bird and moth populations and conducting other such inventories; however, "citizen scientist" still conveys in the other half of this implicit binary the idea of the professional scientist.
77. Waterton, "An Ornithological Letter," 3. This letter is dated 10 March 1837. It is bound together with other published letters and pamphlets at the Linnean Society, London.
78. Swainson was quite proud of his recipe for preserving bird specimens. Writing to Abbot, Swainson boasted of his recipe's effectiveness: "I will put you in possession of a receipt for both [birds and insects] which has enabled me to bring the whole of my collections home in the most perfect state notwithstanding all the destructive little insects that are found in S. America[;] it is indeed invaluable, and was it known might [have] save[d] entire Museums from decay but I believe the secret rests with me." Swainson's secret recipe consists of arsenic, soap, potash, quicklime, and camphor, and he was willing to share it with Abbot because Swainson had asked Abbot for one hundred bird skins for his collection and wanted them to arrive in England undamaged by insect infestation. Swainson warned Abbot to take care when using this recipe: "The least drop falling under the nails, it

will separate the nail from the skin & cause some pain" (Swainson to Abbot, 28 January 1819, Elmgrove, Swainson Correspondence).

79. Waterton, "An Ornithological Letter," 5.
80. Waterton, 5, 14.
81. For naturalists' animus toward natural history illustrators, see Tobin, "Butterflies, Spiders, and Shells."
82. See Calhoun, "From Oak Woods and Swamps," for critiques of Abbot's drawings by French lepidopterists who thought his drawings failed to convey enough information about butterflies' legs and abdomens with his focus primarily on wing patterns.
83. For a photograph of a plant hopper, in this case a spotted lanternfly, see Bug Guide, "*Lycorma delicatula*," https://bugguide.net/node/view/1916873/bgimage.
84. "Upon examining these attentively, I found that they clearly formed two nondescript and very distinct genera belonging to the subfamily of *Fulgorellae*" (Kirby, "Characters of Otocerus and Anotia" [1822], 13). Kirby's article was read at the Linnean Society in 1819 and published in 1822. For more on Kirby's life and work, see Freeman's biography, *Life of the Rev. William Kirby* (1852).
85. Swainson, *Cabinet Cyclopaedia*, 126:99, quoted in Duncan, *Natural History of Exotic Moths*, 70.
86. Duncan, 69–70.
87. Scudder, "John Abbot, the Aurelian," 150. This article is full of factual inaccuracies and questionable assumptions, some corrected by W. F. Kirby (British Museum curator) in a response to this article, "John Abbot, the Aurelian," 230–32.
88. Scudder, 150.
89. Although the International Code of Zoological Nomenclature was published in 1905, taxonomists such as Smith supported the tradition of honoring the principle of priority in the naming species long before the code was institutionalized. For a fascinating analysis of the politics of naming species within 1830s natural history elites, see McOuat, "Cataloguing Power" (2001).
90. Latreille, *Histoire naturelle, Entomologie* (1819), 41; and Duncan, *Natural History of Exotic Moths*, 69.
91. Dos Passos, "The Authorship of the Names Proposed in *The Natural History of the Rarer Lepidopterous Insects of Georgia*" (1958), 191.
92. Dos Passos, 192. For confirmation of dos Passos's conclusions, see Wilkinson, "Smith and Abbot."
93. Natural History Museum, Brian Pitkin, and Paul Jenkins, *Butterflies and Moths of the World* (data set), 2023, https://doi.org/10.5519/s93616qw.

Chapter 7. Abbot's Legacy

1. Wilkinson points out that Cowan and Rogers-Price repeated Dow's erroneous estimation of Abbot's scientific training. See Wilkinson, "John Abbot's London Years. Part 3," 224.
2. Augustus Oemler's letter was published in full in Dow, "John Abbot, of Georgia," 70.
3. This story is related in Dow, "Work and Times of Dr. Harris" (1913), 116–17.
4. Dow, "John Abbot, of Georgia," 68.
5. See Wilkinson's discussion of Abbot's scientific training in "John Abbot's London Years. Part 3," 224. I am indebted to Wilkinson's work on Abbot's early years in London and his careful reconstruction of this period in Abbot's life.
6. Abbot's early work is labeled according to Linnaean nomenclature in his own hand. Harvard has a volume of drawings that Abbot had prepared as if for publication dated 1772, and the Carnegie Museum of Natural History has the Avinoff drawings, some of which are labeled using Linnaean terminology and dated 1767–72.
7. "Professor" here means someone who professes to study insects, not the holder of an academic title. Drury, *Illustrations of Natural History*, 1:xvi. According to Wilkinson, "There were no more avid disciples of Linnaeus in London than Drury and his fellow naturalists. . . . Drury used Linnaean nomenclature in the *Illustrations*" ("Abbot's London Years. Part 3," 222–23).
8. See Drury's notebook held in the Entomology Library of the Natural History Museum. It lists who bought his book, when, and what they paid for it. In 1770 Abbot ordered the "best" copy, meaning one colored by Moses Harris.
9. James Edward Smith to Richard Pulteney, 1 May 1797, GB-110/JES/MS238/17, Linnean Society, London. Oemler met Abbot in 1805, eight years after the publication of *Insects of Georgia* and Smith's letter to Pulteney. For a discussion of possible errors Abbot may have made in *Insects of Georgia*, such as inaccurate depictions of larvae and improper host plants, see Calhoun, "A Glimpse into a 'Flora et Entomologia.'"
10. Abbot to Swainson, 3 August 1819, Savannah, Swainson Correspondence. His modesty is also apparent in another letter to Swainson: In 1818, after his wife had died and he had thoughts of traveling to Tennessee or Florida to collect birds, insects, and plants, he wrote to Swainson asking him where he might go to "employ my small Talents to the best advantage" (Abbot to Swainson, 1 May 1818, Savannah, Swainson Correspondence).
11. Wilkinson, "John Abbot's London Years. Part 3," 225.
12. Wilkinson, 224.
13. Wilkinson, 224. See Cowan, "Boisduval and Le Conte" (1969), 31; and Rogers-Price, "John Abbot in England and North America" (1984), 42. Dow is an unreliable source for information about Abbot's life and accomplishments: Dow stated that Abbot arrived in Georgia in the 1790s and returned to England in the 1810s (67).
14. "Abbot, John, 1751–1840, John Abbot notes and drawings, 1766–1887, Abbot, John, fl. 1797. [Drawings of insects]," Harvard Library Viewer, https://iiif.lib.harvard.edu/manifests/view/drs:444739812$61.
15. Dow, "John Abbot, of Georgia," 68.
16. For the informal learning of botany, see Sarah Easterby Smith, *Cultivating Commerce* (2017).

17. Dow, "John Abbot, of Georgia," 69.
18. Dow, 66. Since much of Dow's article is riddled with errors of fact, I do not believe he accurately represents Smith and Abbot's relationship, for which Dow provides no evidence or citation. We do not know if Smith bought Abbot's drawings and specimens from Francillon, as Dow asserts, though it does seem likely that Francillon was the source of these items, as he was acting as Abbot's agent by the early 1790s. In Smith's papers held in the Linnean Society, there is an introduction to *Insects of Georgia* written by Abbot in which he addresses a "you" who could be either Francillon or Smith on matters of insect taxonomy. Either way, it is clear that Abbot knew that his work was in the process of being published and was aware of how credit might or might not have been distributed evenly between him and the unknown "you." See Calhoun, "From Oak Woods and Swamps," for another take on how Smith came to work on Abbot's drawings, having been asked by the bookseller Edwards to tidy up Abbot's prose and add Linnaean nomenclature. Calhoun in his 2006 article, "A Glimpse into a 'Flora et Entomologia,'" was more inclined to think Abbot knew Smith was working on the publication.
19. Dow, "John Abbot, of Georgia," 66–67.
20. Dow, "Work and Times," 117.
21. Alexander Wilson, the ornithologist who went birding with Abbot, celebrated his work, defending fieldwork over "the closet": "Well authenticated facts deduced from careful observation, precise descriptions, and faithfully pourtrayed representations drawn from living nature, are the only true and substantial materials with which we can ever hope to erect and complete the great superstructure of science; without these all the learned speculations of mere closet theory are but 'the baseless fabricks of a vision'" (Wilson, *American Ornithology* [1808–14], 3:ix). Bucking the hierarchy that placed the abstract over the experiential, Audubon celebrated "the life of an active naturalist," which meant for him days spent on collecting expeditions shooting birds, drawing, and journaling: "Lehman and I take a walk, and Ward, his gun, dog, and basket, returning [to camp] when hungry or fatigued, or both. We draw uninterruptedly till dusk, after which, another walk, then write up journals, and return to rest early." Comparing this life of an active naturalist to the learned one, Audubon contends: "It is nonsense ever to hope to see in the closet what is only to be perceived—as far as the laws, arrangements and beauties of ornithological nature is concerned,—by that devotion of time, opportunities, and action, to which I have consecrated my life, not without hope that science may benefit by my labours" (Audubon to G. W. Featherstonhaugh, 7 December 1831, St. Augustine, East Florida, in Herrick, *Audubon the Naturalist*, 2:12–13). Abbot, even more than Audubon, was a field naturalist in that Abbot never left the field, while Audubon made expeditions to the field as need arose.
22. For "reasoned images," see Daston and Galison, *Objectivity*, 86.
23. Thaddeus W. Harris to John Eatton LeConte, 1830, Cambridge, J. E. Le Conte Correspondence, American Philosophical Society, quoted in Calhoun, "John Abbot's 'Lost' Drawings," 1. Calhoun, a lepidopterist, has done a very thorough job in assessing the accuracy of Abbot's original drawings of butterflies and their host plants for *Insects of Georgia*. For a detailed discussion of the relatively few errors Abbot made in delineating butterflies, their larvae, and their host plants, see Calhoun, "A Glimpse into a 'Flora et Entomologia.'"
24. For the argument that art-making is thinking, see Pamela H. Smith, "Art, Science, and Visual Culture in Early Modern Europe" (2006). See also Smith, Meyers, and Cook, *Ways of Making and Knowing* (2014).
25. Swainson, *Taxidermy*, 99–100.
26. Walckenaer, *Histoire naturelle des insectes*, 1:iv. My translation.
27. For the details of this transaction, see Calhoun, "A Persistent Case of Mistaken Identity," 379. In 2017 Jordan Goodman, historian, was kind enough to tell me about the Parisian spider volume.
28. Abbot to Swainson, 20 December 1816, Savannah, Swainson Correspondence.
29. Swainson to Abbot, 25 October 1817, Bahia Brasil, Swainson Correspondence.
30. Walckenaer consulted plate 6, figure 26 of "Abbot's Georgian Spiders," as the manuscript has been titled by the Muséum national d'Histoire naturelle, and Abbot's notes on page 6: "Prise par Abbot en juin et en juillet sous de vieux bois" (Taken by Abbot in June and July under old wood) (Walckenaer, *Histoire naturelle des insectes*, 1:286).
31. Walckenaer, 1:323. My translation.
32. The number of new species named by Walckenaer is 282, according to Chamberlin and Ivie, "Spiders of the Georgia Region of North America" (1944), 12.
33. Chamberlin and Ivie were working with Francillon's spider volume in London's Natural History Museum, which has 567 spider drawings, while Walckenaer was using a spider volume with 535 drawings. See Chamberlin and Ivie, 12.
34. Chamberlin and Ivie, 12.
35. Chamberlin and Ivie, 5–6.
36. Chamberlin and Ivie, 6.
37. *Catalogue des livres et cartes géographiques de la bibliothèque de feu M. le Baron Walckenaer* (1853), 41.
38. "Et un autre ouvrage d'une haute importance, qui peut-être eût dû être cite le premier (no 513): Abbot, recueil de 535 espèces d'araignées de la Géorgie d'Amérique" (*Catalogue des livres*, ix). The title of Martyn's republication is *Aranei or a Natural History of Spiders, including the principal parts of the well known work of English Spiders by*

Eleazar Albin as also the whole of the celebrated publication on Swedish Spiders by Charles Clerk (London, 1793).

39. The notes for the Hargrett spider album are written on laid paper with a watermark date of 1799 and are in handwriting that is not Abbot's. The small drawings of the Hargrett volume were probably done at the end of the 1790s, when Abbot was working on the drawings for the other two albums. The London album carries the date 1800 in Abbot's handwriting, although the printed title page has 1792 on it. The Paris album has 1798 written in Abbot's handwriting on it. For more information about similarities and differences between the Paris and London albums, see Calhoun, "A Persistent Case of Mistaken Identity."
40. The eighth "Portfolio of Original Figures of Entomological and other Subjects by Le Conte" held at the APS contains 235 drawings of Hemiptera, 126 of Araneina, and 5 of Myriopoda.
41. Hentz, "On North American Spiders" (1832).
42. Hentz, "Descriptions and Figures of the Araneides of the United States" (1842–44). The journal's front cover gives the publication date as 1842, while the title page gives the publication date as 1844, and volume 4 covers the span of 1843–44. Prefacing the article is this announcement: "The Publishing Committee think it proper to inform the readers of this Journal, that the following article is the first of a series on the Araneides of the United States, which has been offered for publication, by the author, to the Boston Society of Natural History. These descriptions and figures will be followed hereafter by others, and the whole will form an illustrated monograph of all the Spiders observed by Professor Hentz in various parts of this country, and will supply a want which has been long felt in the department of our Natural History" (54–55). Segments of a dozen or so pages titled "Continuation" were published in this journal in 1844, 1845, 1846, 1847, and 1850. Scudder was a member of the publishing committee. The monograph Scudder refers to did not materialize as such, and published in its stead was *The Spiders of the United States, a Collection of the arachnological writings of Nicholas Marcellus Hentz* (1875). See McCook, "February 21: Necessity for Revising the Nomenclature of American Spiders" (1888), 75, for the moniker. McCook suggests that because the drawings on which Walckenaer had based his descriptions were "lost" and Hentz had worked from specimens to write his descriptions, American arachnologists felt that Hentz's descriptions were more trustworthy than Walckenaer's. Abbot's arachnid drawings were not lost, and because they resided in Europe in two of the world's greatest natural history museums, American arachnologists found it convenient to ignore them, despite McCook's efforts and Octavius Pickard-Cambridge (1828-1917), an English clergyman and arachnologist, flagging their presence in Paris in 1876 and 1895. For more details about this, see Calhoun's article on Abbot's spider drawings, "A Persistent Case of Mistaken Identity." Hentz had emigrated from France as a young man with his parents in 1816 in the aftermath of the French Revolution and earned his living primarily as a teacher of French in North Carolina, Kentucky, and Alabama. He spent some time as a tutor for a planter's family in 1819 on Sullivan's Island, near Charleston, South Carolina. It is possible that he may have had some contact with Abbot during this time. According to *NCpedia*, in South Carolina "he intensified his entomological studies with a special interest in spiders" (Walser, "Hentz, Nicholas Marcellus"). Hentz wrote to Thaddeus M. Harris on October 24, 1828, "In the South the spiders of a given species are much larger than in the North," a phrase that echoes Abbot's introduction to his spider volumes. On December 19, 1828, Hentz wrote again to Harris, offering to "divide the entomological field with Harris; [I] will specialize on spiders" (Walser).
43. The principle of priority was not codified until 1905 with the International Code of Zoological Nomenclature.
44. McCook, "December 11: The Value of Abbot's Manuscript Drawings of American Spiders" (1888), 431.
45. McCook, "February 21," 77. Chamberlin and Ivie were equally impressed with the quality of Abbot's drawings, "which in our opinion surpass in accuracy as well as artistry the drawings of Hentz" ("Spiders of the Georgia Region," 12).
46. Thorell's letter is printed in McCook, "December 11," 430–31.
47. Chamberlin and Ivie, "Spiders of the Georgia Region," 12.
48. For plummeting insect populations and the dire consequences such losses may have on humans and other animal populations, see Milman, *The Insect Crisis* (2022); and Goulson, *Silent Earth* (2021).

Bibliography

Primary Sources

Abbot, John. "Abbot's Georgian Spiders" (1798). MS 274. Muséum national d'Histoire naturelle, Paris.

Abbot, John. "Aranea" (ca. 1799). Rare Book Folio QL458.42.A7 A2 1799. Hargrett Rare Book and Manuscript Library, University of Georgia Libraries.

Abbot, John. "Drawings of Insects and Plants." MS Typ 426.2. Houghton Library, Harvard University.

Abbot, John. "Drawings of the Insects of Georgia." Francillon Collection. Entomology Library, Natural History Museum, London.

Abbot, John. "Insects of Georgia" (1792). 2 vols. Entomology Library, Natural History Museum, London.

Abbot, John. "Insects of Georgia" (1792–1804). 17 vols. Entomology Artwork SB q A.1c. Entomology Library, Natural History Museum, London.

Abbot, John. "Insects of Georgia" (1827). Stuart A. Rose Manuscript, Archives, and Rare Book Library, Emory University.

Abbot, John. "Insects of Georgia" (1830). MS Typ 426.5. Houghton Library, Harvard University.

Abbot, John. "Moths of Georgia" (ca. 1817–20). Rare Book and Manuscript Library, University of Georgia Libraries.

Abbot, John. "A Natural History of Insects. . . . Drawn and coloured from Nature. Together with a concise and accurate Description" (1772). MS Typ 426.1. Houghton Library, Harvard University.

Abbot, John. "A Natural History of North American Insects." Smith MSS vol. 39, "Varia Scientific." Linnean Society Library, Linnean Society of London.

Abbot, John. "Notes and Observations on the Drawings of the Spiders of Georgia." Vol. 14, Francillon Collection. Entomology Library, Natural History Museum, London.

Abbot, John. "Notes on My Life." In *John Abbot: Birds, Butterflies and Other Wonders*, by Pamela Gilbert. London: Merrell Holberton for the Natural History Museum, 1998.

Abbot, John. "Water Colour Drawings of the Lepidoptera of Georgia." 2 vols. Entomology Artwork SB ff A.7b. Entomology Library, Natural History Museum, London.

Albin, Eleazar. *A Natural History of English Insects. Illustrated with a Hundred Copper Plates, curiously Engraven from the Life: And (for those who desire it) exactly Coloured by the Author Eleazar Albin, Painter. To which are added, Large Notes, and many Curious Observations. By W. Derham, Fellow of the Royal Society.* London, 1724. Reissue of the first edition, 1720.

The Artists Assistant. Philadelphia, 1794.

The Art of Drawing, and Painting in Water-Colours. London: Printed for George Keith, 1757.

Baker, W. B. "John Abbot's Insects of Georgia." *Emory University Quarterly* 15 (1959): 146–53.

Berkenhout, John. *Outlines of the Natural History of Great Britain and Ireland*. London: P. Elmsly, 1769–72.

Boisduval, Jean Alphonse, and Achille Guenée. *Histoire naturelle des insectes; spécies général des lépidoptères*. 7 vols. and atlas. Paris: Roret, 1836–58.

Boisduval, Jean Alphonse, and John Leconte. *Histoire générale et iconographie des lépidoptères et des chenilles de l'Amérique septentrionale*. Paris, 1829–42.

Bullock, William. *Companion to the Liverpool Museum*. London, 1805.

Burgess, Edward, ed. *The Spiders of the United States, a Collection of the Arachnological Writings of Nicholas Marcellus Hentz*. Boston: Boston Society of Natural History, 1875.

Burrell, Rev. J. "A Short Letter to E. Donovan, F.L.S. one of the compilers of Dr. Rees's New Cyclopaedia on the subject of a paragraph in that work, reflecting on the abilities of the members of the Entomological Society." London: William Savage, Printer, n.d.

Catalogue des livres et cartes géographiques de la bibliothèque de feu M. le Baron Walckenaer. Paris: L. Potier, 1853.

Catalogue of the Extensive Collection of Insects . . . from the Museum of George Milne, F.L.S. Sold by Auction by Mr. Thomas, 38 King-Street, Covent Garden. London: Printed by Richard Taylor, 1824.

Collins, David. *An Account of the English Colony in New South Wales*. London: Cadell and Davies, 1798.

Cramer, Pieter. *De uitlandsche kapellen voorkomende in de drie waereld-deelen, Asia, Africa en America*. 8 vols. Amsterdam, 1775–82.

The Delights of Flower Painting. London, 1756.

Donovan, Edward. *An Epitome of the Natural History of the Insects of China*. London, 1798.

Donovan, Edward. *An Epitome of the Natural History of the Insects of India*. London, 1800.

Donovan, Edward. *An Epitome of the Natural History of the Insects of New Holland, New Zealand, New Guinea, Otaheite, and other islands in the Indian, Southern, and Pacific Islands*. London, 1805.

Donovan, Edward. *Instructions for Collecting and Preserving Various Subjects of Natural History*. London, 1794, 1805.

Donovan, E[dward]. *The Natural History of British Insects: Explaining them in their several states, with the periods of their transformations, their food, œconomy, &c. Together with the History of Such Minute Insects as require investigation by the microscope: The whole illustrated by coloured figures, designed and executed from living specimens*. 16 vols. London: Printed for the author and for F. and C. Rivington, 1792–1813.

Drury, Dru. *Directions for Collecting Insects in Foreign Countries*. In *Mammalia*. Vol. 15 of *The Naturalist's Library*, edited by Sir William Jardine. Piccadilly: Chatto and Windus, 1842.

Drury, Dru. *Illustrations of Natural History, wherein are exhibited upwards of two hundred and forty figures of Exotic Insects*. 3 vols. London, 1770, 1773, 1782.

Duncan, James. *The Natural History of Exotic Moths*. In *Entomology*. Vol. 7 of *The Naturalist's Library*, edited by Sir William Jardine. Edinburgh: Lizars, 1841.

Edwards, George. *Essays on Natural History*. London, 1770.

Fabricius, Johann Christian. *Philosophia Entomologica, Sistens Scientiae Fundamenta*. Hamburg and Kiel, 1778.

Fabricius, Johann Christian. *Systema Entomologiae, Sistens Insectorum Classes, Ordines, Genera, Species*. Flensburg and Leipzig, 1775.

Francillon, John. *Description of a Rare Scarabaeus from Potosi, in South America: With Engraved Representations of the Same, Coloured from Nature*. London: C. Whittingham, 1795.

Gray, Asa. "Some North American Botanists: IV. John Eatton LeConte." *Botanical Gazette* 8, no. 4 (1883): 197–99.

Harris, Moses. *The Aurelian: or, Natural History of English Insects; namely Moths and Butterflies. Together with the Plants on which they Feed; A faithful Account of their respective Changes; their usual Haunts when in the winged State; and their standard Names, as given and established by the worthy and ingenious Society of Aurelians. Drawn, engraved and coloured, from the natural Subjects themselves*. London, 1766.

Harris, Moses. *The English Lepidoptera, or, the Aurelian's Pocket Companion: Containing a Catalogue of Upward of Four Hundred Moths and Butterflies*. London: Robson, 1775.

Haworth, Adrian H. *Lepidoptera Britannica: Sistens Digestionem Novam Insectorum Lepidopterorum Quae in Magna Britannia Reperiuntur*. 4 parts in 1 vol. London, 1803.

Haworth, Adrian H. "Review of the Rise and Progress of the Science of Entomology in Great Britain: Chronologically Digested." *Transactions of the Entomological Society of London* 1 (1812): 1–69.

Henkels, S. V. *Catalogue of the Valuable Entomological Library of the late John L. Le Conte, M.D. of Philadelphia: Being the Most Important Library on Entomology ever offered at Public Sale either in Europe or America*. Philadelphia, 1884.

Hentz, Nicholas Marcellus. "Descriptions and Figures of the Araneides of the United States." Read July 1841. *Boston Journal of Natural History, Containing Papers and Communications Read to the Boston Society of Natural History* 4, no. 1 (1842–44): 54–57.

Hentz, Nicholas Marcellus. "On North American Spiders." *American Journal of Science and Arts* 21 (1832): 99–152.

Hunter, Clark, ed. *The Life and Letters of Alexander Wilson*. Vol. 154 of *Memoirs of the American Philosophical Society*. Philadelphia: American Philosophical Society, 1983.

Jardine, Sir William. "Memoir of Dru Drury." In *Mammalia*. Vol. 15 of *The Naturalist's Library*, edited by Sir William Jardine. Piccadilly: Chatto and Windus, 1842.

Jarratt, Devereux. *The Life of the Reverend Devereux Jarratt, Rector of Bath Parish, Dinwiddie County, Virginia, Written by Himself*. Baltimore, Md.: Printed for Warner and Hanna, 1806.

Jones, William. "A New Arrangement of Papilios in a Letter to the President." *Transactions of the Linnean Society of London* 2 (1794): 63–69.

Kirby, W. F. "John Abbot, the Aurelian." *Canadian Entomologist* 20 (1888): 230–32.

Kirby, William. "Ammophila, a new Genus of Insects in the Class Hymenoptera, including the Sphex sabulosa of Linnaeus." *Transactions of the Linnean Society of London* 4 (1798): 195–212.

Kirby, William. "Characters of Otocerus and Anotia." *Transactions of the Linnean Society of London* 13 (1822): 12–23.

Kirby, William, and William Spence. *An Introduction to Entomology: or Elements of the Natural History of Insects*. 2nd ed. 4 vols. London, 1816–26.

Latreille, Pierre André. *Histoire naturelle, Entomologie*. Vol. 9 of *Encyclopédie méthodique par ordre des matières* (Methodical encyclopedia by order of subject matter). 210 vols., 1782–1832. Paris: Agasse, 1819.

LeConte, J. L. "Monograph of the Genus Passimachus." *Annals of the Lyceum of Natural History of New York* 4 (1848): 141–54.

LeConte, J. L., and George H. Horn. *Classification of the Coleoptera of North America*. Smithsonian Miscellaneous Collections 507. Philadelphia: Collins, 1883.

LeConte, John Eatton. *A Monography of the North American Histeroides*. Read September 10, 1844. Academy of Natural Sciences of Philadelphia, 1844, 32–86.

Lettsom, John Coakley. *The Naturalist's and Traveller's Companion, Containing Instructions for Collecting and Preserving Objects of Natural History*. 1772. 2nd ed., London, 1774.

Linnaeus, Carl. *Systema Naturae per Regna Tria Naturae, Secundum Classes, Ordines, Genera, Species cum Characteribus, Differentiis, Synonymis, Locis*. 10th ed., 2 vols. Holmiae [Stockholm], 1758–59.

Marsham, Thomas. *Entomologia Britannica: Sistens Insecta Britanniae Indigena, Secundum Methodum Linnaean Disposita*. Vol. 1, *Coleoptera*. London: J. White, 1802.

Martyn, Thomas. *Psyche: Figures of Nondescript Lepidopterous Insects from different parts of the world*. London, 1797.

McCook, Henry C. "December 11: The Value of Abbot's Manuscript Drawings of American Spiders." *Proceedings of the Academy of Natural Sciences of Philadelphia* 40 (1888): 428–31.

McCook, Henry C. "February 21: Necessity for Revising the Nomenclature of American Spiders." *Proceedings of the Academy of Natural Sciences of Philadelphia* 40 (1888): 74–79.

Merian, Maria Sybilla. *Der Raupen wunderbare Verwandlung und sonderbare Blumennahrung*. Nuremberg, 1679, 1683.

Merian, Maria Sybilla. *Metamorphosis Insectorum Surinamensium*. Amsterdam, 1705.

Nichols, John. *Literary Anecdotes of the Eighteenth Century*. 9 vols. London: Nichols, 1814.

Review of *Insects of Georgia*. *Analytical Review* 27 (January–June 1798): 1–6.

Rösel von Rosenhof, August Johann. *Der monatlich-herausgegebenen Insecten-Belustigung*. Nuremberg, 1746–61.

Samouelle, George. *General Directions for Collecting and Preserving Exotic Insects and Crustacea; Designed for the Use of Residents in Foreign Countries, Travellers, and Gentlemen Going Abroad*. London: Longman, 1826.

Scudder, Samuel H. "Dr. John Lawrence LeConte, a Memoir Read to the National Academy of Science, April 17, 1884." *Transactions of the American Entomological Society* 11 (1884): i–xxvii.

Scudder, Samuel H., ed. *The Entomological Correspondence of Thaddeus William Harris, M.D.* Boston: Boston Society of Natural History, 1869.

Scudder, Samuel H. "John Abbot, the Aurelian." *Canadian Entomologist* 20 (1888): 150–54, 230–32.

Smith, James Edward, and John Abbot. *The Natural History of the Rarer Lepidopterous Insects of Georgia. Including Their Systematic Characters, the Particulars of Their Several Metamorphoses, and the Plants on Which They Feed. Collected from the Observations of Mr John Abbot. Many Years Resident in That Country*. 2 vols. London, 1797.

Smith, Pleasance, ed. *Memoir and Correspondence of the Late Sir James Edward Smith*. 2 vols. London: Longman, 1832.

Swainson, William. *A Preliminary Discourse on the Study of Natural History*. London, 1834.

Swainson, William. *Taxidermy: With the Biography of Zoologists, and Notices of their Works*. London, 1840.

Walckenaer, Charles-Athanase. *Histoire naturelle des insectes: Aptères*. 4 vols. Paris, 1837–47.

Waterton, Charles. "An Ornithological Letter to William Swainson, Esq. F.R.S. &c. &c. by Charles Waterton, Esq. Walton-Hall." Wakefield: Richard Nichols Bookseller, 1837, 3–15.

Wilkes, Benjamin. *One Hundred and Twenty Copper-Plates of English Moths and Butterflies: representing Their Changes into the Caterpillar, Chrysalis, and Fly States, and the Plants, Flowers and Fruits whereon they feed: Coloured with great Exactness from the Subjects themselves: with a Natural History of the Moths and Butterflies Describing the Method of Managing, Preserving, and Feeding them*. 2nd ed. London, 1773.

Wilson, Alexander. *American Ornithology; Or, The Natural History of the Birds of the United States: Illustrated with Plates Engraved and Colored from Original Drawings taken from Nature.* 9 vols. Philadelphia, 1808–14.

Wilson, Alexander. *American Ornithology: Or, The Natural History of the Birds of the United States: Illustrated with Plates Engraved and Colored from Original Drawings taken from Nature.* 3 vols. Edited by George Ord. New York: Collins, 1828–29.

Newspapers

Daily Savannah Republican
Pennsylvania Gazette
Republican; and Savannah Evening Ledger

Secondary Sources

Afrin, Sadia, et al. "Intercropping Empower Reduces Insect Pests and Increases Biodiversity in Agro-Ecosystem." *Agricultural Sciences* 8, no. 10 (2017): 1120–34.

Allen, David Elliston. *The Naturalist in Britain: A Social History.* 1976; Princeton, N.J.: Princeton University Press, 1994.

Allen, Elsa G. "History of American Ornithology Before Audubon." *Transactions of the American Philosophical Society* 41 (1951): 386–591.

Bassett, Anna Stowell. "Some Georgia Records of John Abbot, Naturalist." *The Auk* 55, no. 2 (1938): 244–54.

Beckert, Sven. *Empire of Cotton: A Global History.* New York: Vintage Books, 2014.

Belknap, Geoffrey. "Illustrating Natural History: Images, Periodicals, and the Making of Nineteenth-Century Scientific Communities." *British Journal for the History of Science* 51, no. 3 (2018): 395–422.

Bergson, Henri. *Creative Evolution* (1907). Translated by Arthur Mitchell (1911). Mineola: Dover Publications, 1998.

Bigelow, Allison Margaret. "Gendered Language and the Science of Colonial Silk." *Early American Literature* 49, no. 2 (2014): 271–325.

Bleichmar, Daniela. "The Geography of Observation: Distance and Visibility in Eighteenth-Century Botanical Travel." In *Histories of Scientific Observation*, edited by Lorraine Daston and Elizabeth Lunbeck. Chicago: University of Chicago Press, 2011.

Bleichmar, Daniela. "Painting as Exploration: Visualizing Nature in Eighteenth-Century Colonial Science." *Colonial Latin American Review* 15, no. 1 (2006): 81–104.

Bleichmar, Daniela. "Training the Naturalist's Eye in the Eighteenth Century: Perfect Global Visions and Local Blind Spots." In *Skilled Vision: Between Apprenticeship and Standards*, edited by Cristina Brasseni. London: Berghahn Books, 2007.

Bleichmar, Daniela. *Visible Empire: Botanical Expeditions and Visual Culture in the Hispanic Enlightenment.* Chicago: University of Chicago Press, 2012.

Bleichmar, Daniela. *Visual Voyages: Images of Latin American Nature from Columbus to Darwin.* New Haven, Conn.: Yale University Press, 2017.

Bleichmar, Daniela, and Peter C. Mancall, eds. *Collecting Across Cultures: Material Exchanges in the Early Modern Atlantic World.* Philadelphia: University of Pennsylvania Press, 2011.

Blum, Ann Shelby. *Picturing Nature: American Nineteenth-Century Zoological Illustration.* Princeton, N.J.: Princeton University Press, 1993.

Bragg, William Harris. *De Renne: Three Generations.* Athens: University of Georgia Press, 1999.

Burtt, Edward H., Jr., and William E. Davis Jr. *Alexander Wilson: The Scot Who Founded American Ornithology.* Cambridge, Mass.: Harvard University Press, 2013.

Calhoun, John V. "Butterfly Drawings by John Abbot in the Hargrett Rare Book and Manuscript Library, University of Georgia." *Journal of the Lepidopterists' Society* 61, no. 3 (2007): 125–37.

Calhoun, John V. "Butterfly Drawings by John Abbot in the Houghton Library, Harvard University, That Are Wrongly Attributed to an 'Inferior Copyist.'" *Journal of the Lepidopterists' Society* 61, no. 4 (2007): 228–32.

Calhoun, John V. "From Oak Woods and Swamps: The Butterflies Recorded in Georgia by John Abbot (1751–C.1840) Based on His Drawings and Specimens." *Journal of the Lepidopterists' Society* 73, no. 4 (2019): 211–56.

Calhoun, John V. "A Glimpse into a 'Flora et Entomologia': *The Natural History of the Rarer Lepidopterous Insects of Georgia* by J. E. Smith and J. Abbot (1797)." *Journal of the Lepidopterists' Society* 60, no. 1 (2006): 1–37.

Calhoun, John V. "*Histoire générale et iconographie des lépidoptères et des chenilles de l'Amérique septentrionale* by Boisduval & Le Conte (1829–[1837]): Original Drawings Used for the Engraved Plates and the True Identities of Four Figured Taxa." *Journal of the Lepidopterists' Society* 58, no. 3 (2004): 143–68.

Calhoun, John V. "History of the Lepidoptera Illustrations in the Thomas Cooper Library, University of South Carolina." ResearchGate, April 14, 2004, 1–5. https://www.researchgate.net/publication/266526081.

Calhoun, John V. "John Abbot's Butterfly Drawings for William Swainson, Including Comments About

Abbot's Artistic Methods and Written Observations." *Journal of the Lepidopterists' Society* 61, no. 1 (2007): 1–20.

Calhoun, John V. "John Abbot's 'Lost' Drawings for John E. Le Conte in the American Philosophical Society Library, Philadelphia." *Journal of the Lepidopterists' Society* 60, no. 4 (2006): 211–17.

Calhoun, John V. "An Overlooked 18th Century List of North American Lepidoptera." *News of the Lepidopterists' Society* 52, no. 1 (2010): 3–5, 16–20.

Calhoun, John V. "A Persistent Case of Mistaken Identity: Charles A. Walckenaer's Collection of Spider Drawings by John Abbot Is in Paris, Not London." *Journal of Arachnology* 47 (2019): 377–80.

Calhoun, John V., and David W. Johnson. "John Abbot, the English Naturalist-Artist, in Virginia." *Banisteria* 35 (2010): 3–10.

Carere, C., and J. Mather, eds. *The Welfare of Invertebrate Animals*. Cham, Switzerland: Springer, 2019.

Catalogue of a Fine Collection of Rare Books and Drawings, on Ornithology, Botany, Entomology, and Other Natural History, Voyages and Travel, the Property of a Lady. Sotheby & Co., November 4 and 5, 1963.

Chamberlin, Ralph V., and Wilton Ivie. "Spiders of the Georgia Region of North America." *Bulletin of the University of Utah* 35, no. 9 (1944), Biological Series 8, no. 5:1–267.

Charmantier, Isabelle, and Stefan Müller-Wille, eds. "Worlds of Paper: An Introduction." *Early Science and Medicine* 19, no. 5 (2014): 379–97.

Chico, Tita. *The Experimental Imagination: Literary Knowledge and Science in the British Enlightenment*. Stanford, Calif.: Stanford University Press, 2018.

Chico, Tita. "From Fleece to Fleets; or Wool and the Production of Wonder." *Eighteenth-Century Fiction* 32, no. 1 (2019): 101–21.

Chico, Tita. "Gimcrack's Legacy: Sex, Wealth, and the Theatre of Experimental Philosophy." *Comparative Drama* 42 (2008): 29–49.

Cole, Lucinda. *Imperfect Creatures: Vermin, Literature, and the Sciences of Life, 1600–1740*. Ann Arbor: University of Michigan Press, 2016.

Coleman, Deirdre. *Henry Smeathman, Flycatcher: Natural History, Slavery, and Empire in the Late Eighteenth Century*. Liverpool: Liverpool University Press, 2018.

Coleman, Deirdre, and Robert Blackburn. "Eighteenth-Century West African Insects in the Macleay Museum, University of Sydney." *Archives of Natural History* 44, no. 2 (2017): 356–62.

Collins, David. *An Account of the English Colony in New South Wales*. Vol. 1. London: Cadell and Davies, 1798. http://adc.library.usyd.edu.au/data-2/colacc1.pdf.

Cowan, C. F. "Boisduval and Le Conte: *Histoire générale et iconographie des lépidoptères et des chenilles de l'Amérique septentrionale*." *Journal of the Society for Bibliography of Natural History* 5, no. 2 (1969): 125–34.

Cowan, C. F. "John Francillon, F.L.S., a Few Facts." *Entomologist's Record and Journal of Variation* 98 (1986): 139–43.

Dall, W. H. *Spencer Fullerton Baird, a Biography*. Philadelphia: J. B. Lippincott Company, 1915.

Daston, Lorraine. "Attention and the Value of Nature in the Enlightenment." In *The Moral Authority of Nature*, edited by Lorraine Daston and Fernando Vidal. Chicago: University of Chicago Press, 2003.

Daston, Lorraine. "The Empire of Observation, 1600–1800." In *Histories of Scientific Observation*, edited by Lorraine Daston and Elizabeth Lunbeck. Chicago: University of Chicago Press, 2011.

Daston, Lorraine, and Peter Galison. *Objectivity*. New York: Zone Books, 2007.

Daston, Lorraine, and Elizabeth Lunbeck, eds. *Histories of Scientific Observation*. Chicago: University of Chicago Press, 2011.

Daston, Lorraine, and Katharine Park. *Wonders and the Order of Nature, 1150–1750*. New York: Zone Books, 1998.

Delbourgo, James. *Collecting the World: Hans Sloane and the Origins of the British Museum*. Cambridge, Mass.: Harvard University Press, 2017.

Delbourgo, James. "Divers Things: Collecting the World Under Water." *History of Science* 49 (2011): 149–85.

Delbourgo, James. "Listing People." *Isis: A Journal of the History of Science Society* 103, no. 4 (2012): 735–42.

Delbourgo, James, and Nicholas Dew. *Science and Empire in the Atlantic World*. New York: Routledge, 2008.

De Vos, Paula. "Apothecaries, Artists, and Artisans: Early Industrial Material Culture in the Biological Old Regime." *Journal of Interdisciplinary History* 45, no. 3 (2015): 277–336.

Dos Passos, Cyril F. "The Authorship of the Names Proposed in *The Natural History of the Rarer Lepidopterous Insects of Georgia* (1797)." *Lepidopterists' News* 12, nos. 5–6 (1958): 191–92.

Dos Passos, Cyril F. "The Dates of Publication of the *Histoire générale et iconographie des lépidoptères et des chenilles de l'Amérique septentrionale*, by Boisduval and Le Conte 1829–1833[–1834]." *Journal of the Society for the Bibliography of Natural History* 4, no. 1 (1962): 48–56.

Dos Passos, Cyril F. "The Entomological Reminiscences of William Henry Edwards." *Journal of the New York Entomological Society* 59, no. 3 (1951): 129–86, 149–50.

Douglas, Starr. "The Making of Scientific Knowledge in an Age of Slavery: Henry Smeathman, Sierra Leone and Natural History." *Journal of Colonialism and Colonial History* 9, no. 3 (2008): https://doi.org/10.1353/cch.0.0029.

Douglas, A. Starr, and E. Geoffrey Hancock. "Insect Collecting in Africa During the Eighteenth Century and William Hunter's Collection." *Archives of Natural History* 34, no. 2 (2007): 293–306.

Dow, Robert Percy. "John Abbot, of Georgia." *Journal of the New York Entomological Society* 22 (1914): 65–72.

Dow, Robert Percy. "Work and Times of Dr. Harris." *Bulletin of the Brooklyn Entomological Society* 8, no. 7 (1913): 106–18.

Ehle, John, Jr. *Trail of Tears: The Rise and Fall of the Cherokee Nation*. New York: Anchor Books, 1989.

Eisner, Thomas. *For Love of Insects*. Cambridge, Mass.: Harvard University Press, 2003.

Elsner, John, and Roger Cardinal, eds. *The Cultures of Collecting*. London: Reaktion Books, 1994.

Elwood, Robert W. "Assessing the Potential for Pain in Crustaceans and Other Invertebrates." In *The Welfare of Invertebrate Animals*, edited by C. Carere and J. Mather. Cham, Switzerland: Springer, 2019.

Endersby, Jim. *Imperial Nature: Joseph Hooker and the Practices of Victorian Science*. Chicago: University of Chicago Press, 2008.

Etheridge, Kay. *The Flowering of Ecology: Maria Sibylla Merian's Caterpillar Book*. Leiden: Brill, 2021.

Feeser, Andrea. *Red, White, and Black Make Blue: Indigo in the Fabric of Colonial South Carolina*. Athens: University of Georgia Press, 2013.

Festa, Lynn. *Fiction Without Humanity: Person, Animal, Thing in Early Enlightenment Culture*. Philadelphia: University of Pennsylvania Press, 2019.

Findlen, Paula. *Possessing Nature: Museums, Collecting, and Scientific Culture in Early Modern Italy*. Berkeley: University of California Press, 1994.

Freeman, John. *Life of the Rev. William Kirby, M.A. Rector of Barham*. London: Longman, 1852.

Gascoigne, John. *Science in the Service of Empire: Joseph Banks, the British State and the Uses of Science in the Age of Revolution*. Cambridge: Cambridge University Press, 1998.

Gilbert, Pamela. *John Abbot: Birds, Butterflies and Other Wonders*. London: Merrell Holberton for the Natural History Museum, 1998.

Goodman, Jordan. *Planting the World: Joseph Banks and His Collectors: An Adventurous History of Botany*. New York: William Collins, 2021.

Goulson, Dave. *Silent Earth: Averting the Insect Apocalypse*. New York: Harper, 2021.

Graves, Algernon. *The Society of Artists of Britain, 1760–1791; the Free Society of Artists, 1761–1783; a complete dictionary of contributors and their work from the foundation of the societies to 1791*. London: George Bell, 1907.

Griffin, Francis J., and Howard K. Swann. "A Revised Census of the Known Copies of T. Martyn's Psyche 1797." *Archives of Natural History* 12, no. 2 (1985): 213–17.

Hancock, E. Geoffrey. "John Russell Malloch: Amateur Naturalist to Professional Taxonomist." In *Naturalists in the Field: Collecting, Recording, and Preserving the Natural World from the Fifteenth to the Twenty-First Century*, edited by Arthur MacGregor. Emergence of Natural History. Leiden: Brill, 2018.

Hancock, E. Geoffrey. "The Shaping Role of Johann Christian Fabricius (1745–1808): William Hunter's Insect Collection and Entomology in Eighteenth Century London." In *William Hunter's World: The Art and Science of Eighteenth Century Collecting*, edited by E. G. Hancock, N. Pearce, and M. Campbell. Farnham, England: Ashgate Publishing, 2015.

Hancock, E. Geoffrey, and Georgina Brown. "AnUnusual Preparation of an Eighteenth-Century Spider and Its Consequences." *Journal of the Institute of Conservation* 37, no. 2 (2014): 110–19.

Hancock, E. Geoffrey, Georgina Brown, and Brian Jowett. "Pinned Down: The Role of Pins in the Evolution of Eighteenth Century Museum Insect Collections." *Museum History Journal* 4 (2011): 29–46.

Hellman, Geoffrey T. "Black Tie and Cyanide Jar." *New Yorker*, August 21, 1948, 32–44.

Herrick, Francis Hobart. *Audubon the Naturalist: A History of His Life and Time*. 2 vols. New York: D. Appleton and Company, 1917.

Hindle, Brooke. *The Pursuit of Science in Revolutionary America*. 1956; New York: Norton, 1974.

Hollerbach, Anne Larsen. "Of Sangfroid and Sphinx Moths: Cruelty, Public Relations, and the Growth of Entomology in England, 1800–1840." *Osiris* 11 (1996): 201–20.

Hünniger, Dominik. "Nets, Labels, and Boards: Materiality and Natural History Practices in Continental European Manuals on Insect Collecting 1688–1776." In *Naturalists in the Field: Collecting, Recording, and Preserving the Natural World from the Fifteenth to the Twenty-First Century*, edited by Arthur MacGregor. Emergence of Natural History. Leiden: Brill, 2018.

Iconotypes: A Compendium of Butterflies & Moths; Jones' "Icones" Complete. Introduction by Richard Vane-

Wright in partnership with Oxford University Museum of Natural History. Oakland: University of California Press, 2021.

Impey, Oliver, and Arthur MacGregor, eds. *The Origins of Museums: The Cabinet of Curiosities in Sixteenth- and Seventeenth-Century Europe*. Oxford: Clarendon Press, 1985.

Jardine, Nicholas, James A. Secord, and Emma C. Spary, eds. *Cultures of Natural History*. Cambridge: Cambridge University Press, 1996.

Jorink, Eric. "Between Emblematics and the 'Argument from Design': The Representation of Insects in the Dutch Republic." In vol. 1 of *Early Modern Zoology: The Construction of Animal Science, Literature, and the Visual Arts*, edited by Karl A. E. Enenkel and Paul J. Smith. 2 vols. Brill: Leiden, 2007.

Jovanovic-Kruspel, Stefanie. "The Art of Painting Butterflies." In *Iconotypes: A Compendium of Butterflies & Moths; Jones' "Icones" Complete*. Introduction by Richard Vane-Wright in partnership with Oxford University Museum of Natural History. Oakland: University of California Press, 2021.

Jovanovic-Kruspel, Stefanie, and Harald Bruckner. "Ignaz Schiffermüller and the Colors of Butterflies." In *Insects and Colors Between Art and Natural History*, edited by V. E. Mandrij and Giulia Simonini. Leiden: Brill, 2024.

Kaeppler, Adrienne L. *Holophusicon: The Leverian Museum—an Eighteenth-Century English Institution of Science, Curiosity, and Art*. Altenstadt: ZKF, 2011.

Kareem, Sarah Tindal. *Eighteenth-Century Fiction and the Reinvention of Wonder*. Oxford: Oxford University Press, 2014.

Kauffman, Thomas DaCosta. *The Mastery of Nature: Aspects of Art, Science, and Humanism in the Renaissance*. Princeton, N.J.: Princeton University Press, 1993.

Kemp, Christopher. *The Lost Species: Great Expeditions in the Collections of Natural History Museums*. Chicago: University of Chicago Press, 2017.

Latour, Bruno. "Drawing Things Together." In *Representation in Scientific Practice*, edited by M. Lynch and S. Woolgar. Cambridge, Mass.: MIT Press, 1990.

Latour, Bruno. *Pandora's Hope: Essays on the Reality of Science Studies*. Cambridge, Mass.: Harvard University Press, 1999.

Latour, Bruno. *Science in Action: How to Follow Scientists and Engineers Through Society*. Cambridge, Mass.: Harvard University Press, 1987.

Latour, Bruno. "Visualization and Cognition: Thinking with Eyes and Hands." *Knowledge and Society: Studies in the Sociology of Culture Past and Present* 6 (1986): 1–40.

Leach, William. *Butterfly People: An American Encounter with the Beauty of the World*. New York: Pantheon, 2013.

MacGregor, Arthur. *Curiosity and Enlightenment: Collectors and Collections from the Sixteenth to the Nineteenth Century*. New Haven, Conn.: Yale University Press, 2007.

MacGregor, Arthur, ed. *Naturalists in the Field: Collecting, Recording, and Preserving the Natural World from the Fifteenth to the Twenty-First Century*. Emergence of Natural History. Leiden: Brill, 2018.

Macknight, Elizabeth. *Aristocratic Families in Republican France, 1870–1940*. Manchester: University of Manchester Press, 2012.

Magee, Judith. *The Art and Science of William Bartram*. University Park: Penn State University Press, 2007.

Mandrij, V. E., and Giulia Simonini, eds. *Insects and Colors Between Art and Natural History*. Leiden: Brill, 2024.

Marder, Michael, and Patricia Vieira. "Writing Phytophilia: Philosophers and Poets as Lovers of Plants." *Frame* 26, no. 2 (2013): 37–53.

Marsh, Ben. "The Republic's New Clothes: Making Silk in the Antebellum United States." *Agricultural History* 86, no. 4 (2012): 206–34.

Marsh, Ben. "Silk Hopes in Colonial South Carolina." *Journal of Southern History* 78, no. 4 (2012): 807–54.

McBurney, Henrietta. *Illuminating Natural History: The Art and Science of Mark Catesby*. London: Paul Mellon Centre for Studies in British Art, 2021.

McOuat, Gordon. "Cataloguing Power: Delineating 'Competent Naturalists' and the Meaning of Species in the British Museum." *British Journal for the History of Science* 34, no. 1 (2001): 1–28.

Meyers, Amy R. W., ed. *Art and Science in America: Issues of Representation*. San Marino, Calif.: Huntington Library, 1998.

Meyers, Amy R. W., ed. *Knowing Nature: Art and Science in Philadelphia, 1740–1840*. New Haven, Conn.: Yale University Press, 2012.

Meyers, Amy R. W., and Margaret Beck Pritchard, eds. *Empire's Nature: Mark Catesby's New World Vision*. Chapel Hill: University of North Carolina Press, 1998.

Miller, David Philip. "Joseph Banks, Empire, and 'Centers of Calculation' in Late Hanoverian London." In *Visions of Empire: Voyages, Botany, and Representations of Nature*, edited by David Philip Miller and Peter Hahns Reill. Cambridge: Cambridge University Press, 1996.

Miller, David Philip, and Hans Peter Reill, eds. *Visions of Empire: Voyages, Botany, and Representations of Nature.* Cambridge: Cambridge University Press, 1996.

Miller, Lillian B., ed. *The Selected Papers of Charles Willson Peale.* Vol. 2. New Haven, Conn.: Yale University Press, 1988.

Milman, Oliver. *The Insect Crisis: The Fall of the Tiny Empires That Run the World.* New York: Norton, 2022.

Müller-Wille, Steffan, and Isabelle Charmantier. "Lists as Research Technologies." *Isis: A Journal of the History of Science Society* 103, no. 4 (2012): 743–52.

Murphy, Kathleen S. *Captivity's Collections: Science, Natural History, and the British Transatlantic Slave Trade.* Chapel Hill: University of North Carolina Press, 2023.

Murphy, Kathleen S. "Collecting Slave Traders: James Petiver, Natural History, and the British Slave Trade." *William and Mary Quarterly* 70, no. 4 (2013): 637–70.

Murphy, Kathleen S. "James Petiver's 'Kind Friends' and 'Curious Persons' in the Atlantic World: Commerce, Colonialism and Collecting." *Notes and Records* 74 (2020): 259–74.

Neri, Janice. *The Insect and the Image: Visualizing Nature in Early Modern Europe, 1500–1700.* Minneapolis: University of Minnesota Press, 2011.

Neri, Janice, Tara Nummedal, and John V. Calhoun. *John Abbot & William Swainson: Art, Science, and Commerce in Nineteenth-Century Natural History Illustration.* Tuscaloosa: University of Alabama Press, 2019.

Neville, Richard. *Mr. J. W. Lewin: Painter & Naturalist.* Sydney: State Library of New South Wales, 2012.

Noblett, William. "Dru Drury's *Directions for Collecting Insects in Foreign Countries.*" *Bulletin of the Amateur Entomologists' Society* 44 (1985): 170–78.

Noblett, William. "Dru Drury's Letters (1770–1775) to the Cambridge Bookseller, John Woodyer." *Transactions of the Cambridge Bibliographical Society* 10 (1994): 539–47.

Ogilvie, Brian W. "Attending to Insects: Francis Willughby and John Ray." *Notes and Records of the Royal Society* 66 (2012): 357–72.

Ogilvie, Brian W. "Nature's Bible: Insects in Seventeenth-Century European Art and Science." *Tidsskrift for Kulturforskning* 7, no. 3 (2008): 5–21.

Ogilvie, Brian W. *The Science of Describing: Natural History in Renaissance Europe.* Chicago: University of Chicago Press, 2006.

Parikka, Jussi. *Insect Media: An Archaeology of Animals and Technology.* Minneapolis: University of Minnesota Press, 2010.

Parrish, Susan Scott. *American Curiosity: Cultures of Natural History in the Colonial British Atlantic World.* Chapel Hill: University of North Carolina Press, 2006.

Parsons, Christopher M., and Kathleen S. Murphy. "Ecosystems Under Sail: Specimen Transport in the Eighteenth-Century French and British Atlantics." *Early American Studies* 10, no. 3 (2012): 503–29.

Pomian, Krzysztof. *Collectors and Curiosities: Paris and Venice, 1500–1800.* Cambridge: Cambridge University Press, 1987.

Pressly, Paul M. *On the Rim of the Caribbean: Colonial Georgia and the British Atlantic World.* Athens: University of Georgia Press, 2013.

Prince, Sue Ann, ed. *Stuffing Birds, Pressing Plants, Shaping Knowledge: Natural History in North America, 1730–1860.* Philadelphia: American Philosophical Society, 2003.

Pugliano, Valentina. "Specimen Lists: Artisanal Writing or Natural Historical Paperwork?" *Isis: A Journal of the History of Science Society* 103, no. 4 (2012): 716–26.

Rehn, James A. G. "The John Eatton Leconte Collection of Paintings of Insects, Arachnids, and Myriopods." *Proceedings of the American Philosophical Society* 98 (1954): 442–48.

Rennicks, Rich. "The History of Vellum and Parchment." *The New Antiquarian: From the Blog of the ABAA,* December 29, 2022. https://www.abaa.org/blog/post/the-history-of-vellum-and-parchment.

Ritvo, Harriet. *The Animal Estate: The English and Other Animals in Victorian England.* Cambridge, Mass.: Harvard University Press, 1989.

Rogers-Price, Vivian. "John Abbot in England and North America: His Accomplishments as Artist and Naturalist." *Turnbull Library Record* 17 (1984): 61–80.

Rogers-Price, Vivian. *John Abbot in Georgia: The Vision of a Naturalist Artist (1751–ca. 1840).* Exhibition catalog. Madison, Ga.: Madison-Morgan Cultural Center, 1983.

Salmon, Michael, et al. *The Aurelian Legacy: British Butterflies and Their Collectors.* Berkeley: University of California Press, 2000.

Saunt, Claudio. *A New Order of Things: Property, Power, and the Transformation of the Creek Indians, 1733–1816.* New York: Cambridge University Press, 1999.

Saunt, Claudio. *Unworthy Republic: The Dispossession of Native Americans and the Road to Indian Territory.* New York: Norton, 2020.

Schaffer, Simon, Lissa Roberts, Kapil Raj, and James Delbourgo. *The Brokered World: Go-Betweens and Global Intelligence, 1770–1820.* Science History Publications, 2009.

Secord, Anne. "Corresponding Interests: Artisans and Gentlemen in Nineteenth-Century Natural History." *British Journal for the History of Science* 27, no. 4 (1994): 383–408.

Simpson, Marcus B. "Artistic Sources for John Abbot's Watercolour Drawings of American Birds." *Archives of Natural History* 20 (1993): 197–212.

Simpson, Marcus B. "The Artist-Naturalist John Abbot (1751–ca. 1840): Contributions to the Ornithology of the Southeastern United States." *North Carolina Historical Review* 61, no. 3 (1984): 347–90.

Smith, Pamela H. "Art, Science, and Visual Culture in Early Modern Europe." *Isis: A Journal of the History of Science Society* 97, no. 1 (2006): 83–100.

Smith, Pamela H., and Paula Findlen, eds. *Merchants and Marvels: Commerce, Science, and Art in Early Modern Europe*. London: Routledge, 2002.

Smith, Pamela H., Amy R. W. Meyers, and Harold J. Cook, eds. *Ways of Making and Knowing*. Ann Arbor: University of Michigan Press, 2014.

Smith, Pamela H., and Benjamin Schmidt, eds. *Making Knowledge in Early Modern Europe: Practices, Objects, and Texts, 1400–1800*. Chicago: University of Chicago Press, 2007.

Smith, Sarah Easterby. *Cultivating Commerce: Cultures of Botany in Britain and France, 1760–1815*. Cambridge: Cambridge University Press, 2017.

Spary, E. C. "Scientific Symmetries." *History of Science* 42, no. 1 (2004): 1–46.

Spary, E. C. *Utopia's Gardens: French Natural History from Old Regime to Revolution*. Chicago: University of Chicago Press, 2000.

Stacey, Robyn, and Ashley Hay. *Museum: The Macleays, Their Collections and the Search for Order*. Cambridge: Cambridge University Press, 2007.

Stearns, R. P. *Science in the British Colonies of America*. Champaign: University of Illinois Press, 1970.

Stephens, Pauline Tyson. "The Silk Industry in Georgia." *Georgia Review* 7, no. 1 (1953): 39–49.

Stone, Witmer. "Some Unpublished Letters of Alexander Wilson and John Abbot." *The Auk* 23, no. 4 (1906): 361–68.

Swan, Claudia. "*Ad vivum, naer het leven*, from the Life: Defining a Mode of Representation." *Word & Image* 11, no. 4 (1995): 353–72.

Swann, Marjorie. *Curiosities and Texts: The Culture of Collecting in Early Modern England*. Philadelphia: University of Pennsylvania Press, 2001.

Terrall, Mary. *Catching Nature in the Act: Réaumur and the Practice of Natural History in the Eighteenth Century*. Chicago: University of Chicago Press, 2014.

Terrall, Mary. "Following Insects Around: Tools and Techniques of Natural History in the Eighteenth Century." *British Journal for the History of Science* 43, no. 4 (2010): 573–88.

Terrall, Mary. "Frogs on the Mantlepiece: The Practice of Observation in Daily Life." In *Histories of Scientific Observation*, edited by Lorraine Daston and Elizabeth Lunbeck. Chicago: University of Chicago Press, 2011.

Tobin, Beth Fowkes. "Butterflies, Spiders, and Shells: Coloring Natural History Illustrations in Late Eighteenth-Century Britain." In *The Materiality of Color: The Production, Circulation, and Application of Dyes and Pigments, 1400–1800*, edited by Andrea Feeser, Maureen Daly Goggin, and Beth Fowkes Tobin. Farnham, Surrey: Ashgate Publishing, 2012.

Tobin, Beth Fowkes. "Color, Taxonomy, and Exotic Specimens." In *Insects and Colors Between Art and Natural History*, edited by V. E. Mandrij and Giulia Simonini. Leiden: Brill, 2024.

Tobin, Beth Fowkes. *The Duchess's Shells: Natural History Collecting in the Age of Cook's Voyages*. New Haven and London: Yale University Press for the Paul Mellon Centre, 2014.

Tobin, Beth Fowkes. "Historical Introduction to Special Issue on Critical Insect Studies." *Eighteenth Century: Theory and Interpretation* 65, nos. 1–2 (2024): 9–22.

Tobin, Beth Fowkes. "Imperial Designs: Botanical Illustration and the British Botanic Empire." *Studies in Eighteenth-Century Culture* 25 (1996): 265–92.

Tobin, Beth Fowkes. "Learning to Craft." In *Material Literacy in Eighteenth-Century Britain: A Nation of Makers*, edited by Serena Dyer and Chloe Wigston Smith. London: Bloomsbury, 2020.

Tobin, Beth Fowkes. "A Naturalist's Notebooks." *Studies in Eighteenth-Century Culture* 45 (2016): 131–56.

Tobin, Beth Fowkes. "Revisiting the Virtuoso: Natural History Collectors and Their Passionate Engagement with Nature." In *Environmental Criticism for the Twenty-First Century*, edited by Stephanie LeMenager, Teresa Shewry, and Ken Hiltner. London: Routledge, 2011.

Wale, Matthew. "Editing Entomology: Natural-History Periodicals and the Shaping of Scientific Communities in Nineteenth-Century Britain." *British Journal for the History of Science* 52, no. 3 (2019): 405–23.

Wale, Matthew. *Making Entomologists: How Periodicals Shaped Scientific Communities in Nineteenth-Century Britain*. Pittsburgh, Pa.: University of Pittsburgh Press, 2022.

Walser, Richard. "Hentz, Nicholas Marcellus." *NCpedia*, 1988. *Dictionary of North Carolina Biography*, University of North Carolina Press. https://www.ncpedia.org/biography/hentz-nicholas-marcellus.

Werrett, Simon, ed. "Rethinking Joseph Banks." Special

issue of *Notes and Records: The Royal Society Journal of the History of Science* 73, no. 4 (2019).

Whitley, William T. *Artists and Their Friends in England, 1700–1799*. 2 vols. London: Medici Society, 1928.

Wilkinson, R. S. "English Entomological Methods in the Seventeenth and Eighteenth Centuries." *Entomologist's Record and Journal of Variation* 80 (1968): 193–99.

Wilkinson, R. S. "The History of the Entomological Clap-Net in Great Britain." *Entomologist's Record and Journal of Variation* 90 (1978): 127–32.

Wilkinson, R. S. "John Abbot's London Years. Parts 1–4." *Entomologist's Record and Journal of Variation* 96 (1984): 110–23, 155–76, 222–29, 273–85.

Wilkinson, R. S. "The Rise and Fall of the Pincushion." *Entomologist's Record and Journal of Variation* 87 (1975): 142–46.

Wilkinson, R. S. "Smith and Abbot, *The Natural History of the Rarer Lepidopterous Insects of Georgia* (1797), and Its Authorship and Later History." *Entomologist's Record and Journal of Variation* 93 (1981): 213–18; 94 (1982): 122, 159–60.

Wyatt, Edward A., IV. "Dr. James Greenway, Eighteenth Century Botanist, of Dinwiddie County." In vol. 1 of *Genealogies of Virginia Families: From Tyler's Quarterly and Genealogical Magazine*. Baltimore, Md.: Genealogical Publishing Company, 1981.

Yale, Elizabeth. *Sociable Knowledge: Natural History and the Nation in Early Modern Britain*. Philadelphia: University of Pennsylvania Press, 2016.

Yeo, Richard. *Notebooks, English Virtuosi, and Early Modern Science*. Chicago: University of Chicago Press, 2014.

Index

Page numbers in italics indicate illustrations.